D1001785

In Danger at Sea

Adventures of a New England Fishing Family

Captain Samuel S. Cottle

Down East Books
Camden, Maine

ISBN-10: 0-89272-753-5
ISBN-13: 978-0-89272-753-7

Cover and text design by Chilton Creative

5 4 3 2 1

Printed at Versa Press, Inc. East Peoria, Illinois

Down East Books
Camden, Maine
A division of Down East Enterprise
Book orders: 800-685-7962
www.downeastbooks.com
Distributed to the trade by National Book Network, Inc.

Library of Congress Cataloging-in-Publication Data

Cottle, Samuel S.
In danger at sea : adventures of a New England fishing family / by Samuel S. Cottle.
p. cm.
ISBN-13: 978-0-89272-753-7 (trade hardcover : alk. paper)
1. Fisheries--North Atlantic Ocean--Anecdotes. 2. Fishers--Rhode Island--Anecdotes. 3. Cottle, Samuel S. I. Title.
SH213.2C68 2007
338.3'727091631--dc22

2006030460

There are no roses on a
Sailor's grave
No lilies on
The ocean wave
Their only tributes
As seagulls sweep
Are the tears that fall
When their
Loved ones weep

AUTHOR'S NOTE: This poem appeared on a wreath laid by the British Navy Legion at the seafarers' memorial in Kinsdale, Ireland.

TABLE OF CONTENTS

PROLOGUE

> "Those going down to the sea in the ships,
> Doing business on vast waters
> They are the ones that have seen the works of Jehovah
> And his wonderful works in the depths . . . "
>
> Psalm 107: 23, 24 *New World Translation.*

From before these words were recorded in the Bible, some men looked to the sea for adventure and others for fortune, but most sought to make a living. All who had an ongoing relationship with the sea—and it was often a love/hate relationship—had to develop a respect for the all-powerful, never-forgiving, forever-rolling ocean.

My early blessing was having been born into a family of seafarers. My grandfather had served as a cabin boy on square-rigged sailing ships from the age of eight, much as did his father before him. His career as a fisherman continued up to the Great War, then after a period of service, he returned to his first love, the sea, and remained active until he reached the century mark.

No other individual has had such an impact on my life, and it was all to the good. Because of the privilege of living, working, and growing in an environment touched by his personality, morals, and basic attitude on life, all my decisions were made as I imagined he would make them.

I remember visiting him one cold February day when he was in the Ice House, a huge barn down near his dock. He was then in his late nineties and was standing on a concrete floor mending a fish trap, getting it ready for the following spring. Although he had set up a fifty-five-gallon drum as a stove, it was located on the other side of the building, and he received very little benefit from its cherry-red surface.

He was wearing a flannel shirt and a cardigan sweater, opened its entire length. His hands were bare, as was need for trap mending, and he was working with wet twine. I was dressed in a worsted-wool dress suit, a scarf, gloves, and a heavy winter topcoat. I lasted about ten minutes before I had to leave because of the cold. My grandfather wished me well and continued to mend, completely unmindful of the biting cold. That is just one example of the strong constitution that I had witnessed since my youth.

My family forged within me an appreciation for hard work and a respect for people whom many called "characters," those with a distinctive combination of traits. Some of these you will meet in the stories to follow.

In the early years of my life, commercial fishermen were often looked upon as ne'er-do-wells by the townsfolk inland. It was not long, though, before fishing was recognized as a reputable line of work. For one month, to emphasize the importance of the fisheries, the Point Judith Fishermen's Cooperative paid every fisherman; every dockhand working inside the Co-op; and all providers of fuel, equipment, groceries, etc. with brand-new two-dollar bills. The area around the port was flooded with these bills, and it was not long before the local communities and other nearby environs recognized just how beneficial the fisheries were. It seemed to readjust the thinking of most landlubbers.

All of the characters and incidents referred to in the following stories are real and factual to the best of my recollection. The kinds of fishing mentioned are still being carried on, although with some modern modifications. Long-lining for swordfish, as is being carried on today, was not used in area of the North Atlantic that I fished, and for that I am pleased.

I would not in good conscience be able to harvest two-foot-long swordfish or catch sea turtles and porpoises and allow them to drown. Nor do I favor mile-long gill nets that trap and kill all in their path. I care too much for the creatures of the sea to allow such indiscriminate killing. As in all endeavors of commerce, the "bottom line" is too often the motivating force that pushes the desire for moral uprightness into the background.

The day of navigating with a sextant, a lead line, and a compass is now gone. Today, instead of running your time between buoys on a course laid out on a chart with a parallel rule and dividers, you turn on your global-positioning system (GPS), which is tied into your radar and automatic pilot. Assuming all of the electronic devices continue to work, they will give you an accurate compass course to the next preprogrammed point on your course. The new system is a delight to work with, but I still run my time and course from point to point, just in case.

I have never thought of myself as a male chauvinist, one who shows little regard for women in general. Still, it was not until I put these stories to paper that someone called to my attention that I had really not taken into consideration just how difficult it is for wives of seafarers and, especially in my case, fishermen.

When I had completed the first chapter of this book, wherein I describe the sinking of my first trawler, the *Roberta Dee*, I asked my wife to read it over. I left the room for a while, and when I returned, she was crying as she read. I asked what was wrong, and she said this was the first time that she had learned what had happened during the sinking and the rescue.

At the time of the incident I had merely told her that we had lost the boat; I had failed to give her the details, and that had caused her pain these many years. I don't know why I didn't elaborate about what had happened, but I was probably concerned about protecting her from what I thought was unnecessary worry.

I well realize now (and probably have all along, but didn't express it to her) how very valuable she was, and is, to me. The wives of all men who live with a measure of risk—policemen, firefighters, and most certainly fishermen—deserve a great deal of credit for the success of such men. How could I have gone to sea if my wife had been overly worried about me, or afraid to be alone, or unable to raise the children pretty much by herself, or incapable of running a household without my involvement along the way?

Gloria, my wife of fifty-six years, was most able in caring for things ashore. She brought up, quite admirably, two beautiful children, and she not only handled our household but also did all of the bookkeeping for our boat and crew—all of this without a whisper of complaint. Each year, she heard the report from the government agency that names the most dangerous type of business (fishing has held that most dubious honor since time immemorial), and she knew the risks each time I left the dock. Apparently she steeled herself and never let on to me about her concerns; rather, she immersed herself in the family responsibilities and related activities.

I well know, as do most seamen I worked with, that we would never have been able to engage in the profession that we loved and lived for if our wives had not been encouraging.

One example of the exceptional fortitude and support shown by my wife, especially in our early years, took place prior to our marriage. I had expressed my firm determination to work on the sea and most likely in fishing. Since she, at that time, had no conception of how much this vocation meant to me, I stated very straightforwardly that all decisions I might make in the future were unilateral decisions. She quickly acquiesced, not fully knowing the difficulties that she would have to face in the years ahead.

Most women delight in preparing and caring for their homes—I believe it's called "nesting"—and one important feature of this endeavor is to establish an atmosphere of security. My wife had those same desires, but she did not have the salient sense of security that was so valued by other women. She learned quickly such terms as "water tow" and "broker trip," which in English mean, respectively, "The net came up empty" and "We did not make any money this trip," or "Sorry, but we can't pay the bills this week." In my early days of dragging (trawling), we made what money there was to be earned in the summertime and made every effort to pay the long-standing bills accrued during the winter months. The fishing community was most unusual in that the merchants providing food, oil, and dry goods extended unlimited credit to every fisherman who worked on a "good boat," had a reputation as a hard worker, and was dependable. Stubby Caswell, of Caswell's Market, comes quickly to mind as a kind and generous grocer. Gloria would shop there for our home needs as well as buying "grub" for the boat. When she went to cash out, she would tell Stubby either, "This is for home" or "This is for the boat," and he would thank her profusely for her patronage. Then if other shoppers were nearby, he would quickly say "I'll send the bill" just loudly enough for them to hear, and that would cover Gloria's inability to pay cash.

Stubby was a real gentleman. When the fishing improved—and especially when I was working my own boat—I spoke to him about his many kindnesses and then asked if he had lost much money by extending credit. He said that very few fishermen failed to pay their debts, but he quickly added that these very same people still shopped at his market and made small payments when they could. At that time there was no system of welfare for fishermen, and it was just as well, as it helped men to develop a needed sense of responsibility and kept them from shame. The Asians have a very kind expression in "saving face." Stubby helped many to save face, and I will be ever grateful to him for that.

As I reflect on the growth of the fishing industry along the eastern seaboard, and especially in the area that I fished over some sixty years, I am amazed at the number of boats, as well as their size. The industry has changed greatly, and not all to the good. For example, the only real problems we had to cope with were the weather, the availability of fish, the price on the East Coast, and the capability of our boats and crews.

In addition to the foregoing, captains and owners today must be current in their knowledge of the catch limits allowed for each species, the areas that are off-limits to fishing, and the consequences of catching a small amount of restricted species while working primarily on a legal species. For example, you might be working on grounds that are open for yellowtail flounder, but when you haul back the net you find five hundred pounds of codfish that are totally off-limits. If you keep the cod along with the flounder and move off the grounds to avoid violating the restrictions on cod, you are guilty as sin in the eyes of the law.

Even if you throw the codfish overboard immediately, it's too late. In fact, their fate is sealed the moment they enter your net, because of the air sack in their gut. As soon as the fish are brought to the surface, they are unable to expel the air within them and they die. Here's an idea: Since the cod are already dead and will be wasted if you chuck them overboard, why not gut them, ice them down, and, when you take out your catch, notify some sort of agency that might be called the "Cod Police"?

You don't plan on keeping the fish and you don't intend on selling them, so why not give them to some charity or to homeless people or to some hospital—anywhere—so they are not wasted. Today if you do that, you can be arrested, your boat can be impounded, your crew will lose their jobs, their families will have to go on welfare, and the codfish will end up in some garbage hopper. And the men disposing of the fish had better not take any of them home or they will feel the wrath of the law, as well. It's easy to understand why veteran fishermen have such a difficult time meeting all of the requirements now imposed on them.

Fishing boats have also changed over the years. Most of the fishermen in Galilee used to have small, eighteen- or twenty-foot open-cockpit boats. They would fish around Nebraska Shoal or near the whistle buoy just south of Point Judith, or perhaps at Cox's Ledge or possibly on the East Ground off Old Harbor, Block Island. They seldom ventured any farther. They would use " tub trawls"—stiff, coiled lines with baited hooks spaced at close intervals—and would set them in the early-morning hours. After the last tub of gear was in the water and its end was anchored, they would run back to the beginning of the trawl and begin hauling—by hand. Some would set a pair of lines, after dropping a marker and small anchor on a special spot known only by them and God. They would work this gear all day long.

The only weather that would restrict these fishermen was winter storms, and then the wind had to be "smokin' " to stop them from going out. They had a unique way of working the anchor lines and trawls in freezing weather. Their wives or other loved ones would knit some woolen mittens with only a thumb, and they would be two to three times too large when completed. The knitters would then throw these mittens into boiling water. Of course they would shrink by two or three sizes—to just the right size.

On those below-zero mornings, the fishermen would dip the mittens into seawater, soaking them up to the wrist. Body heat would then be trapped by the tightly woven wool, and hands would stay warm for the rest of the day.

At thirty feet long, my grandfather's trap boat, the *Olive*, was one of the biggest fishing vessels in the harbor. The *Kathy Dick,* owned by Ted and John Dykstra, was thirty feet, as well, and their father owned the *Anna D.* at thirty-two feet. Painted white, she was called the White Elephant by fishermen from Stonington, Connecticut, to Newport, Rhode Island.

It wasn't long, though, before small western-rig draggers (wheelhouse forward) began to be built or bought and added to the Galilee fleet. A decade or so went by, then eastern-rig draggers (wheelhouse aft) began to show up, usually running from fifty to seventy feet in length. With the advent of the eastern rigs, the fleet made longer trips and fished farther offshore than ever before because this design of boat was far more seaworthy.

Harvey Gamage, of South Bristol, Maine, was one of the most

respected boatbuilders on the eastern seaboard at that time. If you had a Gamage boat, you had no trouble getting a crew to go offshore. Now I don't mean to the grounds off Block Island; I mean to the "Gully," or continental shelf. That area was fished primarily from the fall through the spring. It was a twelve- to fifteen-hour steam from Point Judith to reach the "edge," the one-hundred-fathom curve, depending on the area chosen. We worked mostly from Veaches Canyon in the east to Hudson Canyon in the west. We fished in the Gulf Stream, which works its way up from Florida and north along the continental shelf, all the way to Ireland. I've seen it so warm on Veaches in midwinter that the men on deck would be in their shirtsleeves in January, while the boats working inshore of the fifty-fathom curve were complaining that the twine was freezing as soon as they brought the nets on board, and trawl would topple over like a tree cut in the woods rather than collapsing on deck.

If you ran due south from Point Judith, you would get to the continental shelf after steaming a hundred or more miles. It is no place to be during the winter months with a small western rig, although such a boat was right fine if you were fishing within steaming distance of home and could quickly get to safety if a storm made up. When you are fishing the edge at one hundred fathoms, you want a fine sea boat beneath your feet. I remember fishing with Jerry Adams on the *North Wind* one time when the wind was clocked at ninety-eight miles per hour and there was no chance of our making headway toward home. Jerry kept track of our bearings every half-hour or so, and we were averaging six knots sideways. In fact, when the wind finally let go, it took us as long to get back to the edge from deep water as it did to make the shelf from Point Judith at the beginning of our trip. No, the *North Wind* was far from being a Gamage boat, but she did herself right proud in that blow.

Eastern rigs were comfortable to fish in and as safe a boat as you could work on at that period of time. When fishing a western rig and hauling the net back, you often had the trawl swinging over your head as you tried to move all of the fish down into the cod end. When you dropped the net you sure did not want to be under it, especially when it was rough and everything was swinging from one side of the boat to the other. It was the same when you hardened up the catch in the water and were about to take the cod end aboard.

One advantage of eastern rigs was that nothing but twine was ever over your head. Even when you were lifting the loaded cod end out of the water, most everyone could stand clear until it came over the rail and was lowered down inside the deck checkers. Then one man could pull the ropes holding the cod end closed. As soon as they were tripped, that man was almost always able to get clear when the winch man dropped the net on deck.

Even though the industry has now turned to another style of dragger—the big, steel western rig—there will never be a more handsome, nay, beautiful, vessel than a well-built and well-maintained eastern-rig dragger. They had a deep draft, which stabilized the motion of the hull, and they were seakindly, or comfortable in all sea conditions. The majority of eastern rigs were built of wood, so they were warm and quiet and nice to be aboard.

Today the majority of the fleet is steel, although some smaller ones are fiberglass. They are generally of shallow draft and require "flopper stoppers," or "birds," which are small steel plates that are lowered into the water from long steel booms, or outriggers, extending from each side of the vessel once it is away from the dock. As they plane along just below the surface, the flopper stoppers dampen the rolling motion of a shallow-draft fishing boat.

Whatever the vessel, the men walking the deck are seamen. All have a respect, if not a love, for the water and for the work that fishing for a living entails. I hope after reading the stories in the following pages you will be better able to understand the attraction of the ocean and the pleasure and excitement of the catch. For the fisherman always wonders, "What will be brought up from the depths today?"

It is because of the incessant dangers that each man and each vessel face every day they are at the sea that the title was chosen. It is based on Scripture found at 2 Corinthians 11:26: "in dangers at sea . . .," in which the Apostle Paul faced many similar dangers—his for the love of the Christ, ours for the love of our family and the love of the sea.

May you have a fair wind and a following sea as you begin this journey.

ACKNOWLEDGMENTS

Only an immodest person would believe that he or she has been able to attain any degree of success without the generosity and assistance of many along the way.

Dr. J. Stephen Anderson is one who extended his many-faceted acts of kindness. He has been of great encouragement to me during periods of near despair. He arranged for technical assistance, then worked with me week after week as the story developed. He read every segment of the manuscript and made me feel that I just might have a "shot" at getting it to completion. Thanks, Steve.

Daryl Anderson of the South County Museum in Narragansett, Rhode Island, opened to me a treasure trove of photographs of my grandfather and his crew of "trap fishermen," along with some articles relating to local history and diagrams of trap nets drawn by Captain Peter Sprague. We wish Daryl every success as she and the maritime committee present to the public more than one hundred years of living history drawn from the records and recollections of our fishermen.

Jean Ann Pollard taught a creative-writing class that stimulated me to put down on paper the stories that were welling up within me. Thank you, Jean Ann.

Eric Erickson, a Maine artist of national renown, was kind enough to paint the portraits of my two draggers, the *Roberta Dee* and the *Dorothy & Betty II*. I am deeply grateful for his outstanding work.

Comfort and encouragement have been a daily blessing from the wife of my youth—or, as I call her, my child bride. Her love, support, and wise counsel have kept me stable for fifty-six years, and it is my daily prayer that her gifts continue for another half-century. Thank you, "Momma Bear."

Very important to me are those Maine-woods companions who were not only patient through the many hours of my sharing these tales as we sat around the campfire, but even asked for the telling and encouraged my writing them down.

None of these stories could even have begun were it not for the crew of the *Roberta Dee* and the *Dorothy & Betty II*. Of the full complement of men working with me over the years, outstanding in their loyalty were Norman Gilbert, Paul Champlin, Don Morse, and Maurice "Moe" Morse. They started off with me on my own fishing vessels from day one and stayed with me most of the time that I fished for a living. All but Norm Gilbert have "crossed over the bar," but every one will be in my memory until I, too, sail into the sunset. Thank you all.

INTRODUCTION

The sun drops below the western mountains, and dusk wraps itself around the base of Mt. Katahdin in northern Maine, enveloping the shores of Jo-Mary Lake. The wind that freshened just before the sun set now settles into a peaceful embrace of the lake surface. There will be no further contest between water and wind until the new day tomorrow.

An evening chill comes in from the lake, and soon all the campsites are aglow with a variety of fires, ranging from the dainty flicker tended by the two schoolteachers up on the point, to the near conflagration reaching up into the heavens from Captain Sam's, near the very edge of the lake. This is all a prelude to storytelling. Camp chairs are being carried toward the biggest fire, as that blaze foretells a long evening of reliving the exciting experiences of life.

Murray and Bud Mayo begin with some witty tales of the woodsmen who are working the cuttings up on Jo-Mary Mountain. A story about driving a logging truck along the icy ridge of a road on that same mountain thrills us all. Others about snowmobiles or skidders (woods tractors) that broke down deep in the forest as darkness descended add considerable humor to the evening's event.

Then someone asks me to contribute something.
My only woods lore was gleaned while sitting around campfires listening to some of Maine's best storytellers, so I had to recall a few "sea stories," as they were called by those circled around the campfire.

Oftentimes, especially after I had told a particular tale on frequent occasions, I was encouraged to write it down so that others might learn about living and working on the sea.

This is what prompted me to record the adventures that appear in these pages. That, and the fact that I will soon be "crossing the bar," meaning that any stories not set down on paper will be lost to my family and friends.

I hope that you enjoy these adventures as much as I enjoyed living them.

Captain Sam Cottle
Albion, Maine

Captain's Log

Section I
Dangers at Sea

MAYDAY! MAYDAY!

In years past, commercial-fishing vessels were registered with the United States government. Once this happened, usually at a U.S. Customs House, they technically became the property of the government. In the event of a marine casualty (i.e., the sinking of a vessel), the captain/owner was responsible to report in detail how the loss occurred and, in particular, to provide information as to the welfare of the crew. On October 3, 1956, I was obligated to submit the following:

To: Lester W. Raynes
Commander, USCG
Senior Investigating Officer, Marine Inspection
409 Federal Building
Providence, Rhode Island

Re: Oil Screw *Roberta Dee*
Official Number: 250197
Marine Casualty: 09/26/56, 12:54 A.M.
Exact Locality: Loran Bearings IH0 1805; IH4 5900; IH6 1118

Dear Commander Raynes,

In compliance with Coast Guard regulations in reporting a marine casualty, I submit the following report on the loss of the Fishing Vessel *Roberta Dee*.

We had finished our last tow south of Martha's Vineyard, in thirty fathoms of water, and had just hauled in the trawl doors and secured the net and other gear. Then the crew began to put the fish below into the fish hold. I set a course for Point Judith, put the boat on automatic pilot, and went out to help the crew clean up the deck from the last tow's catch.

The wind was beginning to freshen out of the northeast, and visibility at that time was about fifteen to twenty miles. We were full to the hatches, and they were secured for foul weather.

I set a two-hour watch for each of the crew—Norm Gilbert, Maurice Morse, and his brother, Don Morse. Don was to be the last man on watch, and he was to call me when we got to Whistle Buoy R2, just south of Point Judith. I then turned in.

Maurice "Moe" Morse was near the end of his watch, and his brother, Don, was in the wheelhouse with him, finishing up a cup of coffee. As required, Moe was to check everything on deck as well as in the engine room before turning the watch over to the next man. On the *Roberta Dee*, the engine-room companionway was about ten feet aft of the wheelhouse. As Moe went belowdecks, he discovered seawater nearly up to the engine-room floor. The wind was now blowing at about twenty knots out of the northeast, and visibility was down to eight or ten miles.

Don came down to my stateroom and woke me, saying that we were making water "real bad." I immediately went on deck and down to the engine room. Water was now over the floor. We had two 2½-inch pumps on each engine, and those gave us the capacity of pumping ten inches of water in a continuous stream. I had Don check the hose outlets to see if both pumps were working. They were. I then asked Moe to wake Norm, who was sleeping in the fo'c'sle.

While he was doing that, I called the Coast Guard on channel 2182 to advise them of our situation. I provided cross bearings from our loran and told them that I was changing course from our original destination of Point Judith to Block Island, as it was about fifteen miles closer. I was hoping to make Old Harbor on the east side of the island. At that moment I did not think we were sinking, but I was not going to take any chances.

I couldn't believe that a ten-inch stream of water would not keep up with the problem, so I sent Moe and Don into the engine room to check the belts on the pumps, and I also asked them to try to find out where the water was coming from.

All pipes running from the engine's cooling system to the outside hull seacocks were intact. There was no obvious explosion of seawater from under or around the engines, as it would be if the breach in the hull were in the engine room. Moe and Don came back with the report that the great flood of water was coming from the fuel-tank compartment.

The *Roberta Dee* was a converted subchaser built in 1941, at the outbreak of World War II. She had a separate fuel compartment, with a bulkhead aft of the crew's quarters and officer's staterooms, and it contained bronze fuel tanks about ten feet long. These ran from the keel to the deck above and from side to side inside the hull. The bulkhead forward was watertight, as it had remained intact since the time of the vessel's construction. The after bulkhead, though, had been compromised. A section had been removed when the original engines that the navy had installed were taken out. New GM engines were put in their place when the vessel was converted for fishing. A section of the bulkhead had been left out to allow access to the fuel tanks for the new engines. It was from there that the cold Atlantic was pouring into the engine room.

As we steamed toward home that night, we must have struck a submerged object—a log or a tree stump—something big enough to punch a hole in the hull under the tank compartment. With a hold full of fish, and with the sea conditions as they were, we could strike such an object and not feel the impact.

In the meantime, Norman was clearing the dory of stowed gear and nets, releasing the tie-downs and checking to see that the plug was well seated in the bunghole in the after end of the dory's bottom. He brought out life jackets for all of us and dug out some emergency flares from a locker in the wheelhouse. I told the men to put on the life jackets while we had the opportunity.

I put out a mayday call on the emergency channel, as well

as the channel used by fishermen. The fishing boat *Our Gang* out of Stonington, Connecticut, was just off Point Judith, heading in through the east gap in the breakwater when she responded. Sammy Roderick's voice, conditioned by cigarette smoke and whiskey, was as raspy as sandpaper on a schoolhouse blackboard, but it sounded heavenly to me at that moment.

"What the hell's going on, and where you to?"

I gave him my location and asked if he would be willing to put me under tow if he was able to get to us in time. He said he would, that he was swinging off now to intercept me near Block Island, and that he would have his crew rig up a towing cable for us. What I had in mind was this: The *Roberta Dee* was sinking, but if we could get a cable to her before she went down, and could get her under way, there was a very good chance we could keep her going just below the surface until we got into shallow water. In such an event, we would be able to raise her before the winter storms hit.

About that time, Norman ran into the wheelhouse to say that a steamer was heading more or less our way. I told him to get out the flare gun and fire one so it would cross the skipper's line of sight. He did so, and within a few moments we could see a change in the steamer's course. She was apparently coming out of New York, heading east toward the Nantucket Lightship before setting off for Europe. I had Norm fire another flare, and had Moe light a stationary flare and put it on top of the wheelhouse so it could be easily seen. This done, I was sure that the steamer would be alongside in just minutes. Then we saw the ship turn to starboard, returning to her original course!

We couldn't believe it. She was going to ignore a mayday distress call, flares, and our deck lights on, all of which clearly showed that we were a vessel in distress, not some figment of a lookout's imagination. I called every channel on our radio, screaming for assistance—all to no avail. The steamer sailed by us, then beyond us, then out of sight to the east'ard.

There was nothing left to do but concentrate on things that

were within our ability to control. Even though the engine room was rapidly filling with water, both engines kept running, and the generators kept producing electricity for the deck lights, the lights in the wheelhouse, and—our lifeline to help—the radio. As long as those engines were running, we maintained headway toward Block Island, looking for some sign of assistance from the *Our Gang* and/or the Coast Guard.

The Coast Guard Station located in New Harbor, on the west side of Block Island, was in constant contact with us. They had dispatched a "forty-footer," a very tough, steel picket boat, but the crew had to go around Sandy Point on the north end of the island before they could turn into a southeasterly direction toward us. I certainly hoped they could make it in time.

The *Our Gang* was full of fish and was slow making headway in our direction, yet she had the advantage of being able to head in a straight line to our coordinates. At first I was hoping that the *Our Gang* would arrive before the picket boat, as I wanted desperately to save the *Roberta Dee* if at all possible.

The author's ill-fated first vessel, the eighty-three-foot Roberta Dee, *was a converted subchaser built in 1941. The submerged object she struck was never identified.*

But as the situation worsened, I didn't really care who got to us first, as long as they arrived before we went down.

The Coast Guard told me that a plane had been sent to the scene from their New England headquarters in Massachusetts, and that we should be seeing it shortly. This was no sooner said than the ocean around us lit up like New York City.

It was a PV2B just overhead, and the crew had put on their landing lights. They were in radio contact with the picket boat that was on the way, but I was unable to talk with the plane. I did advise the plane crew, via the picket boat, that we were in no need of rubber rafts, as our dory was on deck and ready to be used if need be. They advised us that they would circle overhead "until help arrived." *If it arrives,* I thought to myself.

Suddenly, an engine stopped. As it did, water spewed out of the exhaust pipe. The engine was blown for sure. Now with two pumps out of service, the flooding engine room caused the boat to list to starboard, the side with the heavy gallus frames and drag gear. Things were getting a bit dicey.

I knew that we would have to put the dory overboard, but I didn't want to do so until we could see the lights of either of the two boats heading our way. A dory is a great little sea boat, but that's with one or two men in it. This dory would have to carry four men and would be launched into a stiff northeast wind with rough seas and poor visibility—not a good situation at best, deadly at worst. Even if the plane dropped a raft, there was no way we could get to it given the sea conditions. Our best bet was to stick with the dory.

Next, two events happened within minutes of each other: Norman came into the wheelhouse to tell me that the second engine was nearly covered with seawater; only the air intake was still clear, which meant that it would be stopping soon. Then, Don hollered that he could see some lights just inshore of us.

I had Norm fire off a flare and called into the radio, "Do you see our flare?" The chief petty officer on the picket boat said that they did, and that they would be alongside in ten minutes or so.

I ordered my men into the dory and told them to get it away from the *Roberta Dee* to prevent any possibility of its becoming trapped in the rigging or the fishing gear or the vortex when she went down. When they were safely away, I called the Coast Guard to advise them of the situation. I told them that I was leaving the wheelhouse—and the radio—and would be on the flying bridge.

From the bridge, I watched my crew being picked up by the picket boat and was greatly relieved for them. I had not wanted to be the one to tell their wives that they were lost. Then I began to hear some wicked noise from the one operating engine and knew it would stop momentarily. I also knew that the minute the *Roberta Dee* lost headway, she was bound for Davy Jones's locker, and I was not keen on going with her.

Just as the engine blew, the chief ran the picket boat up the deeply sloped deck of the sinking *Roberta Dee*, right to the flying bridge, and I stepped onto his deck without getting my feet wet.

As soon as I had joined my crew in the cockpit, the chief yelled, "There she goes!"

We turned to look, and the picket boat's spotlight lit up the *Roberta Dee*'s forward deck as my first command started to settle into the ocean. She raised her bow high, with all the remaining air moving forward belowdecks, and the deck plate on the chain locker in the bow blew off with an explosion that threw foamy seawater a hundred feet into the air.

Then she slid quietly, stern first, beneath the heavy seas until she disappeared from view. I will never forget that sight or that feeling. What I had worked so hard to acquire, my very first dragger, had been lost so suddenly.

The chief told us to go below, buckle up the seat belts, and sit tight, as we were in for a rough two-hour ride to Point Judith. I never knew a boat to have seat belts, but after that ride, I definitely knew why this one had them. We were quartering the sea as we headed into port, and every few seconds that little picket boat (and the forty-footer was small for the job she had to do) would become airborne. If we had not been secured with those belts, we would have been smashed against the overhead. When she dove into a trough, the reverse would have taken place. That ride home, as appreciated as it was, took more out of me and my crew than all of the events we had just gone through.

My hat is off to the Coast Guardsmen. Their role as guardians of the seas is no small one, and sadly, it is often little appreciated by "those going down to the sea in ships."

The chief petty officer kindly called the *Our Gang*, which was just coming into sight, to advise Sammy Roderick of the situation. Ours was a sad return to Point Judith, but at least it was a return.

This, Commander Raynes, completes the report to the best of my recollection and ability to record it. Please extend to the men at Block Island and the aircrew in Massachusetts my sincere "Well done."

Respectfully submitted,

Samuel S. Cottle Jr.
Captain, *Roberta Dee*
Galilee, Rhode Island

SAFE PASSAGE HOME

On the night the *Roberta Dee* went down, the northeast wind had been increasing since early afternoon; it had hit about twenty knots when we started to ship water and now, on the way home, it was nearing gale force. In fact, the last two hours in the Coast Guard picket boat were as painful as I had ever experienced aboard any vessel or in any sea conditions. The forty-footer would lift two-thirds of her length out of water as we headed into the increasing seas, then she would dive off the top of the combing waves, nearly standing on her stem. At that point, the twin propellers would break free of the water and whine loudly until she again lifted her bow, putting the screws deep into the next green swell as she strained to get over the oncoming wave.

The chief petty officer and his first mate were strapped into harnesses bolted to the stanchions beneath their seats; this kept their bodies from being smashed against the wheelhouse, or, worse yet, tossed overboard into the cauldron of foam and green seawater boiling up around the picket boat. God, what a beating we were taking! Just as I was about to complain to the chief, I fully recognized the alternative, bit my tongue, and quickly thanked God for the safety brought about by this fine little craft. If not for the picket boat and her crew we would be floating in the dory alongside the *Roberta Dee* as she lay on the bottom of the ocean southeast of Block Island.

"We're nearing the whistle buoy off the Point," the chief hollered down to us. "The seas should ease up shortly."

Once in the lee of the breakwater, the waves subsided and the picket boat headed nearly due west toward the "can" (cylindrical buoy) just inside the west gap; then we turned northward into the breachway between Jerusalem to the west and Galilee to the east. As we neared the docking area, I remembered that my grandfather had named both sides of the breachway years before, when he first settled in Point Judith. I wondered what he would say when I told him about losing the boat. Not much, I guessed; he'd experienced enough at sea that one more loss would not concern him, as long as all hands got off safely. But he did have some money invested in the boat, so he would have some concerns about my immediate future—whether I would be able to find another boat to fish. Everything would pretty much depend on the insurance company.

We slowed down as we entered the breachway and eased in toward the dock of the Point Judith station house. Nearly all of the buildings and boats along both sides of the bulkhead were totally dark.

Normally, at two o'clock in the morning, crews would be climbing aboard their boats, dropping their nets to the deck, and getting them ready to set out. Engines would be coughing themselves into life, and the few Cat engines in the harbor would be rapping loudly as their small booster motors got them started in preparation for the coming day.

But with the nor'easter blowing overhead, no one as yet was easing out of bed. The fishermen would all get a few more hours of sleep before heading to the docks and the coffee shops to talk about the weather, the fishing, and—this morning—about us.

How do you thank a crew that has literally saved your life? There was not much that we could do for the Coast Guardsmen but thank them, commend them, and wish them Godspeed as they headed out again into the storm and New Harbor, on the island. It seems kind of strange, but "thank you" is about all they seemed to want.

Norm, Moe, Don, and I slowly walked up the Coast Guard dock and out onto the road leading up to the Point Judith Dehydration Plant (commonly called the "Trash Plant") and the nearby parking lot where our cars and trucks were. We mumbled a meek "good-night" to each other. I said that I'd call them in a few hours after we'd had a chance to rest and count our blessings.

As I was about to head up the road, Henry Mello hailed me.

"Where's the boat? Down at the Co-op? The Trash Plant wants to shut down as soon as you've unloaded."

"Well, Henry, I lost the *Roberta Dee* southeast of Block Island. The Coast Guard picked us up and just dropped us off at the station house."

"Did you get everyone off okay? No one hurt?"

"We're all okay," I replied. "Scared, but okay."

I had started to pull away when he hollered, "If I can help in any way, let me know. I mean it. Anything."

I waved thanks and drove off. I could hardly keep my eyes open. I didn't realize how tired I was until I sat down in the pickup. "Hope I can get home without having another wreck," I muttered to myself, and headed off on the seven miles to home.

Because I would regularly arrive home at all hours of the night or day, my wife and I never locked our doors. I couldn't remember to carry a key, and if I had, I would most likely have left it on the boat. If Glory or the kids heard any noise in the night, they knew it was just me and would never really wake up. (Of course, those times are past.) I let myself in, went into the bathroom, shaved and showered, and then went into the bedroom.

Just as I was trying to slip into bed quietly, Gloria rolled over, gave me a kiss, and asked, "Have a good trip?"

I responded in as calm a voice as I could muster. "We lost the boat. She sank."

"Oh, that's nice," said Gloria, whereupon she rolled over and was instantly asleep again. That was a big mistake. I've been telling people about her response for the past forty years. She can't get off the hook. She says that she thought I was just kidding. I wish that I had been.

Although I was exhausted, the shower had refreshed me enough to allow me time to reflect on the past day's events before falling asleep. In fact, I began to think about how I had reached this point in my life. I was twenty-one years old; married to my high-school sweetheart; had two beautiful children, Colleen and David; was captain of an eighty-three-foot fishing vessel—or, rather, had been her captain—and was meeting the goals that I had set out for myself years before.

My mother used to delight me with a story about taking me to sea on my grandfather's boat when I was six weeks old. She said that I took to it instantly. As soon as the boat—the *Olive*—sailed out the west gap into the early April sou'west breeze and began to rise and fall gently as she headed into the swell, I apparently fell fast asleep.

As my mother tells it, I stayed asleep until we went back inside the breakwater several hours later. She was certain that I would follow the lead of my grandfather and become a fisherman. Of course, some time passed before I met her expectations completely, but I did take to our seafaring environment quickly and happily.

As I lay in bed that night of the sinking, I tried to bring to mind one of the earliest memories of my youth. I could clearly recall being about five years old and standing by my great-grandfather's side. He was on crutches, and he challenged me to a race to the main house, down by the dock. I accepted the offer and took off as fast as I could. Looking back, it seemed to me that old Cap'n Clark had swung those crutches as fast as he could. But he let me beat him, to my squealing delight. Only years later did I learn that this lively old gentleman had fought in the Civil War. He was fourteen years old and had volunteered as a drummer boy. You may think that was no great feat until you learn that the drummers and flag carriers were in front of the troops as they went into battle. Needless to say, they suffered high casualties.

Old Cap'n Clark began to fish out of Connecticut and Long Island after the war. His son, my grandfather, followed his lead, and after some years of sailing on square-riggers, settled down in Niantic, Connecticut.

Shortly after the turn of the century, my grandfather moved his family and his fish-trap business down east to Point Judith. His was the first big fishing operation in the harbor. As I mentioned earlier, the entrance, or breachway, into Salt Pond was unnamed at that time. In fact, there was not much of a breachway at all.

There was a sandbar between Salt Pond and the area inside the breakwater. This bar would hold back the winter's accumulation of runoff from the land around Salt Pond, as well as any that flowed out of Potter Pond, a brackish body of water lying up behind Matunuck Point. For years the local folks would just dig a small hole in that area of the sandbar, and the buildup of water would gush out, creating a temporary breachway. It would allow limited ingress and egress for a few weeks, but the gap would quickly fill up with silt or sand if hit by a storm.

My grandfather recognized that a more permanent solution was needed if he were to be successful in the trap business, as he needed to go into and out of the pond every day during the fishing season. So he encouraged all of the fishermen and townspeople to help him open up a much larger breachway between two points of high ground, just inside the large stone breakwater offshore. They built a temporary bulkhead across the low-lying spit of beach to hold back the tremendous body of water lying inshore. Then they dug a deep sluiceway just offshore of the bulkhead and blew it out with dynamite. A virtual wall of water, millions of gallons, cascaded out through the newly dug sluiceway, deepening it more. Just a few hours after the bulkhead was removed, a permanent breachway was created.

At a later date, the Army Corps of Engineers inspected the new entrance to Salt Pond. They could see how this would enhance further development of the local fishing industry, so they brought in barges and cranes and millions of tons of huge granite stones. They laid a wall on either side of the new channel, then dredged it even deeper.

It was during all of this activity that my grandfather realized the new development would need to be named, as future nautical charts would make note of this harbor of refuge. He named the west side of the channel Jerusalem and called the east side Galilee. Cap was a student of the Bible and thought these names fitting. They were accepted by the local people, by the Army Corps of Engineers, and eventually by the cartographers; today you will see these names on every nautical chart and map of coastal Rhode Island. My grandfather also named the area in which he made his home, about four miles up-pond from the breachway. He called it Snug Harbor, which is also found on charts and maps of the area.

While a few fishermen lived and worked in and around Galilee, there was very little infrastructure in place. There was one narrow road, which was often covered by an exceptionally high tide, running from the mainland out to the edge of the marsh used by the fishermen. There were no docks, although a handful of men drove down a piling or two just a few feet from the bank. This kept their small craft from drifting under overhanging marsh growth that had been undercut by the tides. Some less-enterprising individuals would come down in the morning, hoping to go fishing for the day, only to find that their boat had eased up lengthwise beneath the marsh edge and that the overhanging topsoil, consisting of sand intertwined with marsh grass and moss, had dropped down into the boat, causing it to sink. This was not an uncommon sight.

My grandfather, on the other hand, immediately laid granite riprap along the frontage abutted by his home. He then topped off and faced the riprap with concrete, which would prevent the tides and waves from undermining the beachfront. Once this was done, Cap rigged up a pile driver on a small barge and began to sink pilings that he had cut and treated with creosote to extend their life in salt water.

He drove two pilings right next to each other, setting another two just opposite these, about ten feet away. He used cables to wrap each pair of pilings together, making a firm foundation for the dock superstructure to follow. Driving more pilings as he went, Cap laid docking all the way out to the deepest part of the channel, which ran pretty well north and south from High Point, up-pond from him, to Galilee to the south'ard. His plan was to develop a marketing center that would be able to handle any and all fish, lobster, and shellfish being caught in and around Point Judith. In no time at all, his dock and unloading area became the center of the local fishing industry.

Several wholesale fish dealers—some from Stonington, Connecticut, others from Providence, and a few retailers such as Mr. Lamb from Westerly—were ever present, buying fish, re-icing them in wooden barrels, and either shipping them to distant markets or selling them locally inland.

Upon reflection, my grandfather was the first entrepreneur in Point Judith. Once he encouraged the local fishermen to sell their catch at his dock and arranged for dealers to buy the seafood, he found that he was earning money on both ends of the fishing business. He then built a long shed on the dock to house local women who would process some of the volume of fish for which there was no immediate market. For example, during the early part of the season, Cap's fish traps would catch thousands of pounds of blowfish—a species of puffer—that had no market value at all. My grandfather had the women cut the fish just behind the head but not all the way through the belly skin; they would then turn the body inside out and pull the tail portion clear of the remaining skin, yielding a portion of clean, snow-white flesh about six or eight inches long, with a single tailbone. Gramp called these bits "Chicken of the Sea," shipped them to New York, and received a prime price for them. In fact, he had a tough time keeping up with the orders.

Another fish he caught in great quantities but had no market for was the sea robin, an ugly creature that was nonetheless delightful in taste and texture. He found a way to sell those, too.

Shortly after putting up the processing shed, Gramp built a fish market, selling fish that he had caught that day, along with some purchased from other fishermen. This establishment attracted people for miles around. Whenever Gramp sold a pound or two of any well-known, accepted species, he would throw in a couple of pounds of cleaned sea robin at no charge. A fillet of robin was just the tail end of the fish, much like the pieces harvested from the blowfish, but there were seven ribs in each fillet. My grandfather would explain to the person buying the fish just how to cook the robin and then, before serving it, how to take a knife and gently lay it alongside the bones and pick them out. The customer then had a wonderful piece of tasty, boneless fish. If Cap's instructions were followed, he or she would return and order sea robin. Before long, my grandfather was able to put a hefty price on it, and folks would buy every fillet available.

Gramp built a second long building right on top of the dock, meaning that it sat over the water. One half of the building, the inshore end, he made into a seafood restaurant that my grandmother ran. My mother and some of my aunts would work with her during the summer months. You cannot believe the meals they put out.

None of my family used recipes—it was a handful of this and a smidge of that—but the end result was heavenly. And the pies! How I would love to have one of my grandmother's pies right now! They were scrumptious.

Gramp moved his fish market into the offshore end of the building. Along its restaurant end, he had two tanks built—one to hold lobsters and the other to accommodate softshell clams and quahogs. He piped in seawater right from the channel, and they all thrived.

A large display case just about in the center of the market's single room was loaded with fresh, clean ice (we also had an ice house on the dock), and a variety of locally caught fish. To the right of the case was a small counter where the fish were weighed and wrapped. On the offshore end of the building there was one long cutting board where we would dress out fish to order or prepare lobsters for broiling or baking. Over the cutting benches was a wall of windows that opened to the channel, allowing the cool sea breeze to flow in on the hottest days.

Here, too, were two or three square chutes into which we would throw the carcasses of the fish we cleaned; they would then drop into the water below. Such remains would draw hungry green crabs to the area; these, in turn, would be caught by me and sold to the city boys, who used them as bait for sea bass and tautog (a Narragansett Indian word for blackfish). Nothing went to waste around my grandfather's place of business.

My grandfather, Cap Clark, was the P. T. Barnum of the seaside. He figured out everything that would draw people to his dock. One of these attractions was fish cars: large rectangular crates with slatted sides and bottoms that allowed a free flow of seawater to enter the interior. The top side of each car was solid-planked but was equipped with large, removable hatches that permitted people to look inside.

In these cars, Cap would imprison sharks of various sizes in one section and codfish, tautog, skates, and any other unique-looking fish in another. (There was a slatted partition between the compartments.) Cap would place benches along the edge of the dock so folks could sit and watch the fish, and he would encourage them to do so. If they got hungry or thirsty, why, there was a restaurant right on the dock; if they needed fish for supper, the market was only steps away.

And if you wanted to fish off the end of the dock, Cap just happened to have a few lines available for rent and some green crabs that you could buy for bait. If you were on the dock early in the morning, he would invite you to go out to watch him haul traps. He would explain that this was going to take a few hours, so you just might need a box lunch, and he knew just the place to buy one—that is, after you bought breakfast. It's interesting that when I was just a boy growing up in that environment, I thought nothing of it, but upon reflection, my grandfather was an exceptionally astute businessman, especially for that time period and that location. In addition, this whirlpool of excitement and activity drew a cast of interesting characters who had a tremendous impact on my young and very malleable intellect.

Section II
The Early Years

Approaching Point Judith from an airplane, one first sees the breakwaters protecting the entrance to the Harbor of Refuge, with Galilee on the right and Jerusalem on the left. Next on the left is Snug Harbor and still farther north is Salt Pond. The arrow indicates Cap Clark's house, which is shown greater detail in the inset.

SNUG HARBOR

My first awareness of each new day at Snug Harbor began with the raucous sound of seagulls, and I awoke to the new opportunity for excitement and adventure as the sun strode into the bunkroom through the huge east window.

How I loved the sound of the gulls as they fought for their fair share of the food exposed by the early ebb tide. I could watch them as they flew by the window, wheeling and rolling into tight turns and dives. I can easily call this new day to mind despite these many years.

Gramps would call to me to breakfast, and I would leap out of bed because he had promised my favorite—pancakes. How funny it is to think back on how much I loved his pancakes. I would stuff down as many as my belly could hold. I knew that this day, like all the ones to follow, would be long and hot and tiring as we hauled the fish traps lying off Point Judith.

After breakfast, Gramps—Cap—and I would walk down to the dock and climb aboard the *Olive*. She was named after my grandfather's first wife—he was to outlive five. My grandmother was the fourth, and at this time was still living. The boat measured thirty feet long and about fifteen feet wide, and she drew about five feet of water. Cap would go down into the engine room, open the fuel line, and slowly turn the flywheel over until its handle was on top and dead center. If it was off center by just an inch or two and Cap spun the wheel to get the engine to fire, the flywheel would snap back like a bolt of lightning, and God bless the one who had his hand near it, as he could end up with a broken arm or worse. But Cap was an expert in running that old Palmer engine. He would talk to it as if it were his first wife, and more often than not, the engine would fire up and begin purring like a big cat. (She was too ornery to be a kitten.)

By the time the old *Olive* was running, most of the crew members were climbing aboard. Some began to move the work skiffs around the end of the dock; they would be used in hauling the fish traps. Other crewmen were preparing to let go the dock lines, and the rest were getting the scoop nets, fish picks, and knives ready to use. Cap would put the *Olive* into forward gear and ease her bow into the dock. The lines would be released, then he would put her into reverse and back her out into deeper water and catch the tide. Once again, the engine was put into forward gear, and with any luck and no little skill, the *Olive* would swing to the east'ard and down the channel, headed out to sea.

Sometimes the strangest creatures showed up in Cap Clark's fish traps. Here a very young Sam Cottle poses with a thresher shark, or "whiptail," as the fishermen called it (for obvious reasons).

I was just a little guy at the time, but Cap had taught me how to steer the boat. I was so short that I had to stand straddled across two lockers placed on either side of the narrow wheelhouse. By doing that I could look out the wheelhouse windows and was able to guide the *Olive* down the channel.

I always got nervous as we neared the State Pier on the Jerusalem side of the harbor, just before we reached the jetty leading out to the main channel alongside the west wall of the breakwater.

There was most always a strange, eddying current that swirled around the pier, and it would either draw the *Olive* in toward the pilings or shunt it way out to the Galilee side of the harbor, toward the boat traffic that was coming and going on that more active side. Cap would quietly step into the wheelhouse from his work on deck and just watch how I handled the wheel. He would not take it out of my hands unless I got too nervous and asked him to. If I appeared to be having trouble, he would speak softly to me as he told me to turn the wheel this way or that, but he wanted me to get the feel of the helm and how the boat reacted to it. He said that was the only way I would learn.

In all of the many years of being around my grandfather, I don't ever remember him yelling or slamming around when things didn't go right as so many others I had seen did regularly. Occasionally I saw him angry, sometimes at me, but he simply begin a dry whistle, *whissss*; it wasn't musical and, as I think back, it sounded more like a radiator losing steam. Even that rare occurrence was short-lived.

Cap had unbelievable patience with people—that is, except with my brother, Johnny. Johnny had gotten into so much trouble over the years (he was four years older than I) that as soon as he showed up, my grandfather would begin his steam whistle. I think that the last straw, so to speak, was when Johnny set fire to one of my grandfather's buildings. I think he was trying to smoke for the first time—I'm not sure—but whatever it was, Johnny was sent down to my Aunt Della's in Galilee.

Once the *Olive* passed the State Pier, we slid by the lobster boats of Gus Damascus and Fish-Fish, which were tied next to each other. I never did know Fish-Fish's real name, but then again, quite a few characters around the harbor were known only by their nicknames.

The next dock was the berth of Captain Steadman and his two sons. They, too, lobstered most of the time. Then came Skip Streeter's dock. He had a big one that accommodated his own boat and, on its inshore end, a fish market. In fact, my grandfather would stop at Skip's on the way back in after hauling the traps and offload any fish that Skip might need for his market.

Beyond Skip's dock, there were only a few summer homes up on the

Two skiffs positioned along the edges of the "parlor" (outermost enclosure) of a fish trap have closed the funnel-like "wings" and are beginning to haul the net in which the catch is confined. If you look carefully, you'll see the general shape of the trap, defined by the floats that keep its upper edge at the surface.

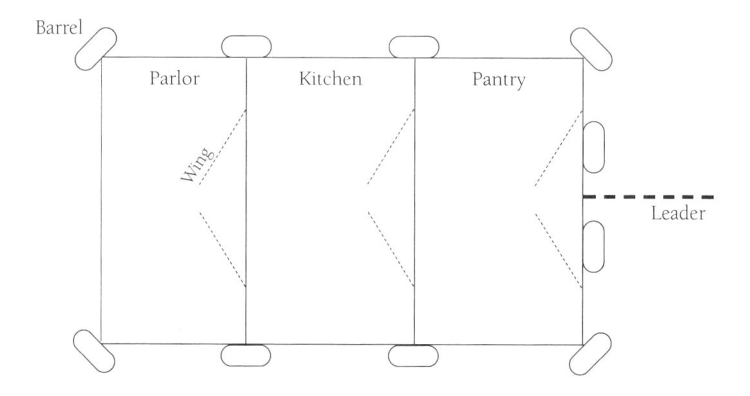

An overhead diagram of the typical fish trap used along the Rhode Island coast shows the "leader," running from the trap to the shore; the anchored barrels used to hold the netting in position and to define the shape of the trap; and the three "rooms," or enclosures, each with a set of closable "wings" designed to funnel the fish ever deeper into the trap.

sand dunes. Once past them, we went out by the narrow jetty, sometimes with great speed and sometimes with great effort, depending on whether the tide was flooding or ebbing. That stretch was a little tricky, because we would be towing four large skiffs and they added to the difficulty of steering. At this point, my grandfather would take the wheel for the rest of the way out to the traps.

Cap held licenses from the state of Rhode Island to place nine fish traps along the shore. He set two east of Point Judith, one at Narragansett Beach, and one near Black Point. A fourth trap was set on the inshore end of the east wall of the Point Judith breakwater, another on the offshore side of the bend in the breakwater, and two more along the offshore side of the west wall. The rest of them were set along Matunuck Beach, all the way down to Moonstone Beach.

Each fish trap was a unique arrangement of nets. If you can imagine the floor plan of a rather small cottage of three rooms, you can see the layout of a typical fish trap. My grandfather's varied in size, but two of the "rooms" probably averaged thirty feet square, with the inshore room measuring about thirty feet by fifteen feet.

Once Cap made the decision as to the location of a trap, he and his crew would bring boatloads of special gear to that location. The first thing to be set was a series of huge iron anchors, which were placed very carefully in relation to each other and to the Harbor of Refuge seawalls. Attached to these anchors were heavy ropes that rose to the surface and were attached to huge, floating barrels. These were designed and built much like wine barrels, but they were much bigger in size and of greater strength in construction. As the trap barrels were set and attached to their respective anchors, other lines were connected from barrel to barrel until three nearly perfect, connected squares were established. From the center of the third, smaller, inshore section, a heavy rope—the leader—

At this point, most of the net in the "parlor" has been hauled in, and the men in the work skiffs are "tightening up" the catch prior to picking up dip nets and "bailing" the fish onto the deck of the Olive*, visible at the right-hand edge of the photograph.*

The author's grandfather, Captain J. E. "Cap" Clark, almost always worked alone in his trap skiff and could outhaul and outlast men half his age. Note the big, anchored trap barrels in the background.

was rowed into the seawall, where another big anchor was set into the rocks. A second heavy rope would be run offshore from the trap for about three hundred feet and secured in a straight line by several large, anchored trap barrels.

Once this framework was established, then the netting would be attached to the lines at the surface (the trap frame) and dropped to the sea floor, turning the three sections into "rooms." The first of these (the one closest to shore) was called the "pantry," then came the "kitchen," and, finally, the "parlor" (at the offshore end of the trap). These were designed to capture and hold fish that hit the leader twine and headed offshore, toward the trap proper, passing through a narrow gap between each pair of "wings," which served as a funnel at the entrance to each room. (See p. 42.)

After the gear had "soaked" overnight, Cap would steam out in the morning and slowly maneuver the *Olive* to the offshore side of the trap. The crew would secure her fore and aft along the line connecting the barrels at the outermost corners of the parlor, then climb into the different trap skiffs they had towed out. Two skiffs with two men in each would go around to the wing on one side of the kitchen, while Cap went alone to the other side.

When in place, and on Cap's signal, the crew would quickly close up the wings to prevent the loss of any fish that might not have moved into the parlor section. Then, working hand over hand, they would pull the twine up and inward, holding a straight line as they moved and allowing the twine just hauled to slide over the skiff and go back overboard. This way, they were not piling up twine that would later have to be reset.

Once the wings and the kitchen itself were hauled, the wings to the parlor were closed; this time, whatever additional twine was hauled stayed inside the work skiffs. Gradually, the body of fish was "hardened up" (pressed together) as tightly as possible. The work skiffs—sometimes two, sometimes as many as five—were then tied together in a line to prevent the fish from finding any kind of opening.

Crew members would begin "bailing" the fish onto the deck of the *Olive* using scoop nets that held a bushel or more each time they were dipped. It didn't take long for this kind of work to tire a man out. Yet, alone in his skiff, my grandfather would bail right along with the younger men, scoop for scoop, and he would outlast them all.

Sometimes there were so many fish in the net that after the *Olive* was loaded down and all of the work skiffs were full, another section of twine, called a "pound," was secured outside the main body of the trap. The balance of the fish remaining in the parlor would be bailed into this holding pen until we went back to port and unloaded the *Olive*; then we would head back out to haul the overflow in the pound. I have seen times when the fish were running so heavily that we had to go out late in the afternoon to empty the same trap a third time.

Literally, tons upon tons of marketable fish would be taken each and every day during the whole summer season, year after year. Today you will find a virtual wasteland in the same territory where Cap used to set his traps; these waters are devoid of all the species he once caught so readily.

Years ago, when I was a boy, there were great schools of mackerel, scup, squid, and sea bass, along with a great variety of other sea creatures that followed the shoreline and the schools of bait that gathered there. We would see sturgeon, sharks of all varieties, and even an occasional turtle. These would be released without harm back into the ocean, although a few interesting individuals might be moved into the floating exhibit at my grandfather's dock, where they would amaze the curious.

Shown here being towed behind the much-larger* Olive*, heavy, wooden skiffs like these were critical to the setting and tending of Cap Clark's fish traps. Note the fish on the* Olive's *deck, confined by removable boards called "checkers."

AN IRISH PRINCESS

I remember my grandmother as a "woman of size." Of course, as a little boy, *everyone* was "of size" to me. She had a wicker rocking chair that sat against the wall in the southwest corner of the kitchen. On that same wall was a shelf that protruded about a foot or so above her head, and on that shelf was a small radio with an elliptical shape rising from its base; it had a maple veneer and a round dial with a dull, yellowish light that barely allowed you to see the different stations as you sought them out. That didn't really matter, though, because the only station my grandmother listened to was the one broadcasting Arthur Godfrey.

My fondest memory is standing by her side as she rocked and wrapping my two arms around her right arm as she rested in her chair. Sometimes I would attempt to climb into her lap, often without success. She had no real lap and filled that chair to the brim, but I would try, and she would help me. Because she did, I knew that she really wanted me closer, but the laws of physics kept me from reaching our mutual goal.

The author describes his grandmother, shown here holding him when he was an infant, as strong, loving, and very influential in his life.

She was a gentle Irish lady, quick of wit and hardworking, as was the wont of that breed and that land. Her maiden name was Geoghegan, and her family came from Castletown Geoghegan in County Westmeath. She often told a story about her family being of the landed gentry and of great means, with castles and the like, but that tale was only told when she was in her cups, if you will.

Like generations before her, my grandmother enjoyed a pint or two upon occasion, which usually involved a family gathering. She would lower her voice to a soft whisper and then tell of the Irish princess who ran away with the coachman that she truly loved.

The young woman's father, of course, disowned her from kith and kin, and she was disinherited from the great wealth that would one day have been hers but for her sin.

"One day, lad, you will come into great wealth and be recognized as a man of stature in Ireland," my grandmother would softly tell me with a loving smile.

Naturally no one in the family believed for a moment that there was a grain of truth in her story, but we did love to hear her tell it because she became caught up in it, and revisiting the possibility of coming out from under the poverty that she had endured most of her life gave her great joy. It was only in recent years, during a trip to my grandmother's homeland and mine, that I came to realize that her history of the family was true—or almost all of it. As for the part about the princess, we will have to wait and see.

This lady, this Princess of West Meath, my grandmother, married an Irish lad named Frank Courtney from Waterville, County Kerry, Ireland. It is said that he was stately, with a thatch of jet-black hair and eyes as blue as the water of Ballinskelligs Bay. This happy union brought about my mother, Genevieve; my aunt, Mae; and my uncle, Sonny. Yet "time and unforeseen occurrence" brought unexpected woe upon the family, which had emigrated to America.

Living in Bangor, Maine, near the end of a trolley line, my mother and her sister would wait for their father to come home from work each evening while sitting in the large front-room window overlooking the street and the tracks of the trolley.

On this never-forgotten day, Frank was hanging onto the steel pole used to assist people onto and off the trolley car. It was accepted custom that the trolley would not stop for those about to get off but would merely slow down. Frank in his youthful arrogance did not wait this day for the slowing of the trolley; instead, he swung out and away, and quickly hit the paved road on the heels of both feet.

The jolt stunned him for a moment, but he did not fall. However, he walked slowly up the sidewalk by the house and passed in front of his daughters without the customary wave and smile. The children ran to their mother, yelling, "Father's not coming into the house, he's gone into the shed. We think he's sick because he didn't wave to us. Let's go find him!"

"You stay in the house, Jenny," said Grammy. "Watch supper and see that it doesn't burn. I'll be right back."

Well, she didn't come right back. When she went into the shed, my grandmother found her first love, her true love, lying quietly in the corner of the shed, propped up against the wall and dying. His skin was turning black, and he was unable to move or breathe. He whispered that the moment he struck the ground he knew he was going to die, and he did not want that to happen in front of the children. Before a doctor could be called, he was dead.

Life was hard on Irish immigrants. Work was nearly impossible to find in Maine at that time, so the princess moved her family to the mill town of Woonsocket, Rhode Island, where she met and married John Girard, a French Catholic. With the blessing of the Church, she bore him fourteen children. One was delivered in the morning, but she rose to provide a noon meal anyway, and when that was over she returned to her job at the mill, carrying the newborn in a rucksack strapped to her back. Eight of those children died of whatever disease was going around; names were seldom given to the cause: apoplexy or the "scours" sufficed for most deaths.

Then Girard himself died, leaving his Irish princess with my mother, Aunt Mae, Uncle Sonny, and six of his offspring. She had no money, no home, no one to care for the children while she worked, and no hope for the future. Forced by circumstance, she put the children into the St. Vincent de Paul orphanage in Providence.

Like the Rachel of antiquity, this lovely lady was "beautiful in form and beautiful of countenance," with long, flowing, auburn hair, yet she suffered with anguish over the loss of her children. She sought employment that would provide better circumstances for them, as her constant thought was to remove them from the orphanage and provide a home for them, one that would allow them to enjoy more of life than they had since Frank's death so many years before.

Her search brought her, by chance, to Snug Harbor, where she worked as a housekeeper, cook, and companion to a wealthy, retired businessman named Chef Wilbur. Chef owned a beautiful little summer cottage on a hill overlooking Salt Pond to the east and Galilee and the ocean to the south. In fact, one could observe the entire breakwater, which ran from Point Judith to Jerusalem, and Block Island, fifteen miles offshore to the southwest. A stone wall surrounded the property, and my grandmother planted pink roses that literally covered it in summer. They so impressed me as a child that roses are my favorite flower to this day.

Although it was not possible to take all the children into this small home, Grammy did take a few at a time to enjoy several days away from the orphanage. By this system of rotation, all got to be with her, especially during the summer months, when they could wander down to the pond and the beach for swimming, fishing, and picking up shells. She kept ever present before them her efforts toward, and fervent desire for, the day when she would have them all with her.

One day my mother, Genevieve, and my aunt Mae wandered the beach toward the home of Captain James Everett Clark, a seaman of some renown in the fishing community around them. J. E., or Cap, depending on who was speaking to him, was a widower, having lost three wives over the years, and he was the father of several sons and daughters. The sons were nearly grown and were involved in the fishing industry with their father; the daughters were married and on their own.

On that day, Cap was out on the beach repairing nets for the fish traps that he had set in various places around the breakwater—two near Narragansett Beach and a couple down toward Matunuck. The girls were laughing and teasing each other, as young ladies will do, especially since they were enjoying a day of freedom from the overbearing orphanage with its strict rules and unloving oversight. Cap stopped his work and watched as they approached. He enjoyed the sound of their laughter. It had been some years since there was joy in his house, and he realized just then how much he missed it.

When the girls reached him, Cap called out and asked who they were and where they were staying. It was a few weeks before the summer tourists would be arriving, and he was curious about the new family in the neighborhood. He had sold most of the land for a few miles around and didn't remember meeting these two young ladies. My mother, the oldest and then the bravest, spoke right up and pretty well told him the family's history. Cap's interest was stirred when Mom spoke of my grandmother, and he decided that it was high time he paid a neighborly call on Chef.

This was done and, as the saying goes, "The rest is history." Before too much time went by, my grandmother and Cap were married, all of the children were brought into the family unit, and all of them fell in love with J. E., or as I called him, Gramps. He was without doubt the best thing that happened to my grandmother since her marriage to the lad from County Kerry.

Years passed. My mother, now grown, married an Englishman, of all things. Not long afterward, my sister, Grace; my brother, John; and I came on the scene. I grew to know and dearly love the warm and gentle woman who no longer looked like the Rachel of old with auburn locks that flowed. Now my grandmother's hair was close-cropped and her form was ample. Her heart was full of love, as was her new home.

In the early years of the author's fishing career, the docks at Galilee saw plenty of activity, most of it involving "day boats," small draggers, fish-trap boats, and lobster boats that offloaded their catches each night.

A MAN OF CHARACTER

When seafarers speak of a "character," they are most often referring to a specific idiosyncrasy displayed by a particular person, and this is usually a diminishing feature. Such was not the case when men along the bulkhead in Point Judith or Snug Harbor spoke of Cap'n Clark.

J. E. Clark, my grandfather, was respected because he was a "man's man." He could be as tough as needed when overseeing a fishing crew of thirty crusty, old "sea dogs," or as gentle as possible when listening to the head of a neighborhood family who needed food for his wife and many children. In addition, Cap loved to laugh. His sense of humor was renowned, and he leaned toward practical jokes.

Captain J. E. "Cap" Clark was an innovative, enterprising businessman, as well as a skilled, hardworking fisherman. The author, his grandson, remembers him as kind, generous, unprejudiced, and fair in his dealings with neighbors and crewmen alike.

There were few days during the summer months when we sailed out of Snug Harbor to haul our fish traps and didn't have ten or twelve "city people" on board. Most of them were folks who rented cottages for the season, sometimes from Cap. Some were "regulars," as Cap called them, because the same families showed up year after year. A few had been spending vacation time in Snug Harbor for three generations. Some of them had purchased land from Cap when they were newly married.

Because many young couples start out "not having two nickels to rub together," as my grandfather said, it was well known that he had practically given them house lots "just to get them started." These self-same folks all loved their homes, Snug Harbor, and the unique individuals in that community.

One of the favorite tricks Cap used to play on unwary visitors would take place on the way in from the traps. The *Olive* would be loaded down with several tons of fish, and oftentimes the work skiffs trailing behind were as full as we could make them and still keep them afloat in the seas that made up around the breakwater.

As soon as we rounded the end of the wall, be it at the West Gap or the East Gap, the *Olive* would settle down in the calm water and Cap would send me or some other crewman down into the engine room. He would give the order loud enough so all the visitors aboard could hear it: "Jack, run down into the engine room and check the bilge water. She's getting logy and difficult to steer! Sam, check the transom. See if it's full of water!"

"The transom is dry Gramps," I would report.

"No water in the engine room, Cap," Jack Lewis would yell. Both of us knew what would happen then.

Cap would look out ahead to see that there was no traffic and to ensure that the *Olive* was being held on course by the becket (loop of rope) that he had set on the spoked wheel just before he began this whole charade. He would then come out on deck with a look of concern on his face. "Well, there is only one other place to check. Sam, open up the fish hold and we'll take a look," he would say fretfully.

Just aft of the engine room and behind a watertight bulkhead was a small tank belowdecks that would hold several hundred pounds of live fish or lobsters in about five hundred gallons of seawater.

In this section of the hull, a number of one-inch-diameter holes had been drilled through the bottom of the boat, allowing the free flow of oxygenated ocean water within the isolated compartment. When we found some unique species of sea creature in our traps—or a healthy-looking shark or the like—in they would go until we returned to our dock in Snug Harbor. They would then be transferred into the large holding car (slatted wooden crate) tied along the north side of the finger pier that was the center of Cap's entrepreneurial universe.

With great theatrics, Cap would lift the cover off this intentionally flooded compartment and shout, "Oh, my God! Look at that water! We're sinking! Quick, Jack, get the hand pump! See if we can keep ahead of the water long enough to make it inside the breachway."

Jack and I or some other crewman would struggle to get the pump out of the junk alongside the wheelhouse and stumble to put it down into the fish well. One of us would start to pump as fast as we could, taking turns as we tired. When we had all tried pumping the unpumpable, Cap would turn to the visitors aboard and in a frantic voice ask them to share in the pumping, urging them, "Keep us afloat or we'll all be lost!" Of course, everyone took his turn and would pump with all his might in this never-ending task.

The pump we used must have come off the Ark, as it was the most antiquated thing that I have ever seen, then or since. It was about six feet long and cylindrical, with a snout near the top through which water would gush. At the lower end of the pump was a leather flap that allowed water to enter. A wooden, shaftlike handle that projected straight up from the main tube was supposed to be the salvation of all aboard.

Had everybody used soda straws, we probably would have had more success in sucking that live well dry, if that had truly been our intent.

During this heroic effort by the "day sailors," Cap kept the *Olive* on course, through the breachway, past Skip Streeter's dock and fish market, by the Jerusalem State Pier, and on up the channel toward home. Of course no one noticed our headway once the pumping began, as Cap kept everyone in a frenzy, constantly checking the "progress" being made by the men on the pumps and telling the crew to "get out the life jackets, bring the work skiffs up close to the stern, be ready to beach in the event we lose the battle." The visitors would then surge ahead with even greater energy in their assigned task as lifesavers.

On the west side of the channel, running north from Jerusalem toward Snug Harbor, there was an area of beach that Cap used off and on during the summer months to haul our trap nets out to dry. As they sat in the water, sea moss accumulated on the twine until the nets had the appearance of shaggy rugs and would weigh so much that the barrels and corks holding the nets up in the water would begin to sink below the surface, allowing fish to stream over the top.

Cap would normally make a wide turn east, then head due west and run the *Olive* ashore on the soft beach sand until the old girl would go farther. We would then haul our twine up onto the sand and go to work cleaning and repairing. When we had finished and had loaded the nets onto the deck of the *Olive*, my grandfather would back the thirty-footer off the beach and head home.

During our staged "sinkings," Cap would basically go through the same motions, but as soon as the bow of the *Olive* ran up on the sand, the water would run out of the fish tank.

Cap would come out of the wheelhouse, peer down into the drained hold and praise the men for their valiant efforts to save the boat and crew. He would haul the pump out of the tank, then close the cover, go back into the wheelhouse, and throw the boat into reverse, sliding her into the channel. He would then put her into forward and head the short distance to his dock.

All the visitors, though weary from their labors, would climb onto the dock with the look of victorious warriors after a battle to preserve home, hearth, and country. No doubt the stories of that day spread throughout the land and kept each and every one of them excited about next summer and the glorious adventures to come.

Over the years, I had a wonderful opportunity to observe the great flood of people that came to Cap's door or sailed with us on the *Olive* or just sat on the benches around the dock. There were actors from a true "Theater by the Sea," people who found an oasis of peace and quiet in the company of J. E. and some of the old barnacles working for him. There were great numbers of dysfunctional families from the towns and cities around us. They arrived, pouring out of their cars, all screaming at each other, the children running from the parents and the parents trying to avoid the responsibility of keeping their children in order.

Cap would come out the front door of his house, calling softly to the children, reaching out to them as they came, embracing each and every one, assigning each of them to work with a crew member. He would ask the wives and mothers to sit on deck ahead of and around the wheelhouse, encouraging them to relax and enjoy the trip. He assigned different tasks to the fathers and husbands, emphasizing the importance of each job and explaining how it would benefit the well-being of all aboard.

As young as I was at the time, I could quickly see that these unhappy, disruptive, and nonproductive families were soon molded into a smooth, cohesive part of the crew. By the time the day's fishing was over, they left the *Olive* in a different frame of mind than when they had arrived. A few hours of Cap's encouraging oversight brought peace into their lives. Such townspeople often returned year after year, and Cap never seemed to tire of their presence.

My grandfather naturally had a strong influence on my personality and the development of my character over the years. My father was a racist, and had I not been able to observe Cap, I might have become one, too. Over the years of my development I never saw one action or heard one word which indicated that Cap was even aware of the difference in the races or in the color of their skin.

One family in particular comes to mind. George was about five feet, nine inches tall with a barrel-like chest and powerful arms. He had sparse, close-cropped hair with ringlets wherever the hair was allowed to grow. He stood erect in an almost military stance, and his face always exhibited the most wonderful smile, exposing brilliant white teeth. One outstanding trait, and probably the one that attracted my grandfather, was his wonderful sense of humor.

Mabel, George's wife, was different in every way except for her humor. She was perhaps the most rotund woman I had ever seen. Her most prominent feature was her breasts. Without exaggeration, they hung below her stomach. As a young boy I had never observed such a phenomenon before and, truthfully, never have since.

The story goes that she had delivered a dozen children or more—I never was able to get a straight answer as to the correct number when I asked. Yet my grandmother had delivered *nineteen* children, eight surviving to maturity, while still retaining a very handsome form and face until her senior years.

It was obvious to all that George's family was poor. Their clothing was tattered and worn yet clean, as were their bodies. Each child had been scrubbed and combed; although barefoot, they exuded happiness and love. Those who could fit in the backseat of their old Pontiac would tumble out of the car and race ahead of their parents down to the dock, where they would peer over the edge into Cap's huge floating car and wonder at its variety of fish. As they swam around in circles, each child would try to name them, calling out excitedly to their dad: "Come and look!" They would cling closely together, protecting each other from falling, and they would stare for as long as their dad would allow.

For George, Mabel, and their children, home was on Sand Turn Road in Kingston, near what is now the University of Rhode Island. A number of families lived near them in shanties situated along both sides of the road. All were of color, all were poor, and all seemed to enjoy the same happy spirit of life that George displayed. They would work at anything that came along: digging potatoes in Charlestown, picking fruit and vegetables in season, helping on the sod trucks in Exeter. Some had jobs at the college, landscaping and building stone walls—most any tasks requiring hard labor. Work—and, therefore, money—was often scarce, so they appreciated any kindness extended.

My grandfather was more aware of this situation than I, and he was also sensitive to their pride and dignity. He was always careful to offer them great quantities of seafood for the work they accomplished—mechanical repairs or construction by the men, or cleaning, cooking, or sewing by the women. (To mask his generosity, he might say that he was about to lose barrels of fish due to the market's being closed before he could ship his catch.) George had a real knack for repairing automobile engines, and whenever he showed up at the dock, Cap always had an engine or two that needed tuning. In return, he would fill the trunk of George's old Pontiac to its limit with mackerel, scup, tautog, or any surplus from the fish market. He would also top off the gas tanks of every car used by George and his friends.

I well remember the time my grandfather started a fish route. My aunt Grace had married Alfie Boudouin, the first of seven husbands. He was a young man from Maine—up in potato country, I think. Alfie had no job, no money, and no place to live. My grandfather offered him all three, but Cap expected this newlywed to put in a day's work for a day's wage.

Alfie was up at dawn to haul traps and when that was done, Cap had him load five barrels into a pickup truck, place block ice and seawater into each, and then fill two barrels with scup and three with beautiful, large mackerel. The barrels were then "headed up" with waterproof tarps and a steel hoop that was driven down over the tarp. This kept the sun off the fish and kept the seawater from sloshing out during travel. My job was to accompany Alfie and, once in the chosen neighborhood, ride in the open truck body to scoop out the fish ordered by the housewives.

The price was three mackerel for a quarter, and the scup were two cents apiece.

Our specific instructions from Gramp were to end each day at George's house on Sand Turn Road. We were to offload any remaining fish, providing George, Mabel, and their children with a feast of fresh, well-iced fish.

In return for this ample gift of food, Cap had asked George and his neighbors to advertise by word of mouth the fine quality and reasonable price of these fresh-caught fish. In George's mind this was food for labor—an honorable exchange. Cap had no idea how successful George and his friends would be, but within a few weeks Alfie and I were making two or three trips to cover our assigned and ever-expanding territory. In fact, we had to hold back from selling out each day so we would have a supply of fish for George.

As I write this I can clearly recall how cold that seawater was with the chilled fish sitting in salt water on those blocks of ice. I was so short at the time that I would have to lift myself up somewhat and lean deeply into the barrel to reach the last of the fish, and I would come up with arms that seemed to be turning blue. I can still feel the chill, but I just loved that job. I was proud of the quality of the fish, and I knew the price was really low, so that most everyone in the "rurals" could afford them. I delighted in seeing the happy and appreciative faces of the folks on Sand Turn Road.

I was especially proud of how my grandfather helped these folks without demeaning them in any way. J. E. Clark heeded the written word to "clothe yourselves with the tender affections of compassion, kindness, and love. . . ." I only hope that I have been successful in following his pattern of life.

I know that I've tried.

THE ITALIAN SCULPTOR

Daylight and a flood tide would be the time to make your first observation of a most unusual man. The old salts hanging around Cap's dock said that this picturesque individual, commonly called George the Guinea, had a most interesting background. Of course, whether you choose to believe the scuttlebutt is up to you.

As the story goes, George (last name unknown) was a promising young sculptor whose skill was causing a stir among the artists in and around Rome. It was just about the time that World War I was beginning, and Italy was about to become involved. A call for national conscription went out, and the army most likely would have inducted George had he not taken drastic steps. Apparently the only solution to the upcoming problem was to escape from the country, but an Italian national could not legally leave during that period.

What George did to successfully remove himself from his homeland remains a complete mystery, but he eventually showed up in Snug Harbor. More correctly, he lived on the south side of the channel running between Potter Pond and Salt Pond, to the east. There the brackish water from Potter Pond mixed with the more saline ocean water in Salt Pond, and that is where Snug Harbor nestled into the sandy shore of the nearby estuaries. It was also the best vantage point from which to observe the comings and goings of this once-promising, hopeful modern Michelangelo.

The older fishermen such as my grandfather said that when George first appeared on the scene, he had the face and physique of what old history books describe as a Roman Centurion. As they told it, when he was "in his cups" he would explode into classical Italian opera and extend his wonderful, warm personality to everyone around him.

But the years and the vino and perhaps the ever-present fear of being found out by the local *polizia* took its toll on the Roman clam digger. And a clever clam digger he was, as he took the flood tide to lessen the burden of rowing to the flats in the upper part of Salt Pond and rode the ebb tide to ease his way home after six hours of bending over a short-handled clam rake.

It was always a delight for me to watch George as he rowed back home, staying close to our dock in order to avoid any head current coming out of Potter Pond. If he nudged up to the edge of the marsh, he could make relatively good headway, while anyone who was inexperienced in the ways of tidal flow and who stayed out in the deeper channel would make little progress, even with an outboard motor. Such people were always amazed to see how George would outdistance their powerboats.

What I enjoyed most about the appearance of George on the fair tide was to contemplate the unbelievable mound of flotsam collected during his journey up Salt Pond. You have to make a clear mental picture of this comedic scene. George was about five foot, eight inches tall, and no longer like the legionnaire of his youth but with a gut hanging eight inches over the belt line (but without a belt), the legs of his pants rolled up to just below the knee, usually one pant leg longer than the other, wearing a tank top undershirt with the bottom edge unavoidably exposing a massive stomach developed by gallons of wine and tons of spaghetti.

You would think that the daily rowing and the agonizing bending over the clam flats for hours at a time would have prevented any expansion of George's body, but the wine must have killed any metabolism that he once had.

This burdensome bit of humanity would stand upright in a skiff no longer than ten feet. At the beginning of the day he looked like an out-of-shape Venetian gondola oarsman, facing the stern of the boat and standing erect while rowing slowly into a favorable tide. On his feet were the most tattered sneakers that you could ever find, and there were always a couple of inches of unbailed water in the bottom of the skiff.

George would eventually arrive "home," to be generous with the word, where he would offload his valuable find of driftwood, glass floats from somebody's net, bits of tattered clothing lost to the sea from some yacht, and any other precious items that had been exposed by low tide. You'd think that he would have separated his treasure into various piles according to an imagined value, but no—George would stack everything in a helter-skelter jumble, much as he'd found it, and transport it down the pond. Then, relieved of this load, he would row back downstream to my grandfather's dock to sell whatever soft-shell clams, littlenecks, and quahogs he had dug during that tide, as well as a few blue crabs, which used to be abundant in the area.

When George climbed onto the dock with his haul, he would invariably lug along his partially empty gallon of red wine, swigging from it as he worked and walked and talked. I don't ever remember seeing him set it down, at least around us. He always seemed a bit tipsy but never downright drunk; still, he was anathema to my grandmother. George was so dirty, slovenly, and uncouth that my grandmother would cringe whenever he walked by the always-open door of our house.

I remember one Thanksgiving Day when the wind was snapping out of the nor'west and the weather was bitterly cold for so early in the winter season. Our family of a dozen or more had just sat down to eat; Grammy had heaped great amounts of delicious-looking food onto plates on the kitchen table, and she had just placed the crispy, golden brown turkey on a platter in front of Cap. He was about to carve when George walked in with his gallon jug of wine, dressed in his usual finery, not showing a hint of being cold. Cap graciously invited him to join us for dinner. If looks could kill, my grandfather would never have lived to be a hundred and two. Grammy was so angry that she was unable to speak, but then words were unnecessary.

I must say that George appeared to sense that the situation was awkward (that is, to everyone but Cap), so he just reached over the table, grabbed one leg of the turkey, gave it a yank while balancing himself with his jug in the opposing hand, and ripped off the entire thigh along with the leg. He then slurred what I took to be a thank-you and headed back into the cold, biting wind before Grammy could take a butcher knife to him. I feared for Gramps's welfare for the balance of that day, though.

In 1938, Rhode Island was hit with the greatest hurricane since the 1800s, and one casualty of the tidal wave that came exploding over the breakwater in Galilee was George the Guinea's home, that mass of derelict timbers set upon a collection of pilings sunk into the tidal flats of Potter's Pond. Fortunately, our Italian boatman was not at home when the wall of water struck. There was nothing left of his ramshackle estate but a few pilings twisted in every direction. George vacated that location and rebuilt a shack on the western shore of Great Island, just across Salt Pond from my grandfather's dock. Having only a modicum of pity for this loss, my grandmother was not unhappy about his forced change of address.

After this turbulent event we saw less and less of the unusual clam digger. Age and excess wine had begun to take their toll on his health. Shortly before his sojourn on this earth ended, nature struck one final blow. We were experiencing an unusually violent thunderstorm that was moving west to east. My grandmother and I were sitting on the front porch of our house, less than fifty feet from the water, and we were watching the brilliant yet frightening display of lightning as it struck along the shore. One bolt hit right in front of us, cracked as loud as a cannon, and flew in a nearly perfect, straight line due east across Salt Pond, toward George the Guinea's shack. There was an explosion of flames, and in minutes the shack was ash.

George never recovered from this "punishment from heaven for leaving Italy and his duty," as he called it. No one could change his mind about what happened. It was not long thereafter that he died.

Daylight and a flood tide would no longer be the time to find that most unusual man, yet he will ever live in my mind and heart as a delightful experience of my youth.

THE BOOT

When I reflect on my childhood, I realize that a prime feature of life is our exposure to the characters who add so much flavor to our existence. One such individual who fascinated me as a boy in Snug Harbor was Pussy Foot Gardiner. I have no idea why or how he earned that moniker, but upon reflection, it fit him perfectly.

My grandmother had just set breakfast on the table and we were about to eat when Pussy Foot walked into the kitchen. Since our door was always open to anyone passing by, this was no surprise. Cap invited Pussy Foot to sit down to breakfast, and as he extended the invitation, my grandmother began to turn fire-red. I thought she was about to slide into apoplexy. Whenever you got within a fathom of Pussy Foot, you had to hold your breath because he smelled bad enough to gag a maggot. He did not know the meaning of the word "bath."

But such a thing never, and I mean never, bothered my grandfather, nor would he ever decline to invite someone to join us at the table.

Pussy Foot limped over and appeared to be in considerable discomfort, but his attention was on the platters of bacon, eggs, and the greatest home fries on earth. He knew my grandmother's talent as a cook, and he was not about to miss out on this opportunity. As Pussy jammed one forkful after another into his mouth, he muttered something about his foot hurting "pretty bad."

Cap said, "When you finish your breakfast, I'll run you up to the hospital. Take your time, though." My grandmother almost choked on "take your time" but held her tongue. Gramps asked me if I wanted to come along, and I jumped at the chance. I was curious as to the reception Pussy would receive in the emergency room.

He was wearing knee boots; in fact, I had never seen Pussy without knee boots, even when he had not been near seawater for days. The staff doctor said to him, "Sit down and we'll take a look at the problem." What amazed me was that the physician didn't even seem to smell the "eau de fish" that floated around Pussy as he obeyed. That was not true of the two nurses who had been in the ER when we first went arrived. They gagged and ran into another room as quickly as possible.

The doctor asked Pussy to remove the boot from the leg that was giving him trouble, and Pussy tried to do so. Having no success, he asked Cap to give him a hand. Now, Cap was six feet tall and had been wrestling fish-trap barrels, iron anchors, and barrels of fish all his life, so he was no weakling. Still, try as he might, he could not get that boot off Pussy's leg. So the doctor decided to cut the boot off, and as he got close to the patient with a knife of some sort, I thought Pussy would hit him.

"Those boots are less than a year old! You ain't going to cut one off! No sir, no sir!"

"Pussy, either you let me cut that boot off, or you're going to have to live with the pain. And it's going to get a lot worse than it is now," said the doctor. Without further hesitation, he grabbed the boot and began cutting, while Cap held the leg up off the floor.

Now, this you aren't going to believe, but it's the truth.

As the doctor got down around the ankle, there arose the most ungodly stench you can imagine. I've smelled a basket of clams left out in the sun and lobster bait crawling with maggots, and the bloated body of a fisherman found floating on a hot summer day. None of these prepared me—or Cap or the doctor—for the putrid, gaslike odor that overwhelmed us as that boot fell to the ground.

"My God, it's gangrene!" said the doctor. "That foot has got to come off! Right now!"

What he was looking at was a black foot and about eight inches of black going up the leg.

Cap said, "Let me take a look at that," and as he did so he picked up a wet cloth and gave the leg one swipe. That took off enough dirt so you could see Pussy had not washed that foot since he had put the boot on it nearly a year before. The doctor got a bucket of warm, soapy water and gently eased Pussy's foot into it, washing as he went. The water quickly blackened and had to be replaced. Eventually the foot was clean enough to examine. The doctor lifted it onto a low stool, slowly rotating it until he saw an ugly, round wound about the size of a quarter.

He then injected Pussy's foot with a needle to kill the pain, and within just a few minutes he had plucked a beer-bottle cap from the wound. It had apparently fallen into the boot, sharp edge up, during one of the social gatherings that Pussy had attended about a week before. He had been walking on that cap ever since and hadn't even taken the boot off to check out what was wrong. As infection took hold, his foot had swollen, causing swelling and the need to cut the boot off.

After Pussy got a tetanus shot, a modest bandage, and Cap's promise to buy him a new set of boots, we were off to the docks. It had just been another typical day at Snug Harbor.

A LONESOME MAN

Looking back, those wonderful, colorful, sometimes rascally characters that permeated my youth while I was growing up in Snug Harbor were not really lonesome men. Each was a loner—"one that avoids others." Having grown up in the middle of a family that experienced all of the passions of life—from love, concern, anger, and frustration to helpfulness and comfort, I could not believe that a person could or would divorce himself from that life-giving environment. Yet as I ponder my relationship with one of those old curmudgeons, Jack Lewis, I appreciate now that he was the personification of a loner, and that being a loner wasn't necessarily bad.

Jack had been a crewman of my grandfather's from my earliest memory, and one that I did not particularly like. He appeared to be sullen, with a countenance like a sou'east squall about to descend on some unsuspecting sailor. Greeted by my grandfather and grandmother when he arrived with the rest of the crew for breakfast, he would just tip his salt-encrusted cap to my grandfather but would remove it when addressing my grandmother and mutter a lowly, "Ma'am." He never spoke to another soul unless addressed directly; then he would respond with the fewest words possible.

Cap thought the world of him as a member of his crew and was quick to say so. It would seem that Jack knew what needed to be done and would do it quickly, correctly, and with a manner of seamanship that just delighted my grandfather. In fact, I do believe that Cap gave him the unspoken assignment of watching over me, especially when I was aboard the trap boat, *Olive*, or when I was in the work skiffs and hauling the fish traps.

I guess I was not the most coordinated deckhand in Cap's crew, having fallen overboard a number of times in circumstances that did not enhance my odds of survival.

A couple of years passed before I became aware that Jack would give some advice to me when I was having trouble in mending the net or splicing a headrope or some other aspect of my work. He would reach over when we were side by side, hauling the twine of the fish trap into the work skiff and the weight was especially heavy from sea moss, and he would help me haul the section I was assigned to. One day, after we had unloaded the day's catch, made all the needed repairs to the gear, and refueled the *Olive* for the next day, Jack told me to get into Cap's work skiff with him. It wasn't really an invitation, it was a command, but one that was given with a more gentle tone than was his nature to display.

"Okay, son, untie the skiff and shove off."

Son? Did I hear him right? *Son?* I quickly did as instructed and we were instantly drifting out in the ebb tide.

"Now pick up the oar and scull back toward the dock."

I did so as I had been taught years ago, and began my struggle to keep the oar in the water long enough to make headway. I had little success.

"Hand me the oar. Now sit down on that side." Jack was sitting on the starboard side of the skiff's transom, and as I swung the oar to him he took it gently into his hand and told me, "Watch the blade of the oar as I move it." As he said that, he cleanly cut the water with the blade and turned his wrist ever so slightly to the left, then to the right. "Don't fight the oar; let it work for you. Just put the edge into the water, move your wrist, and the water pressure will dig the blade down. It will hold the oar from pushing up against you and keep it from popping out of the oarlock."

Far from being a lock of any kind, the oarlock was merely a semicircular cut in the top edge of the transom. It was just a round groove in which the oar was laid as you began to scull. Water pressure held the oar in its place. That is, once you got the hang of it. Jack handed the oar back to me and, following his instructions, I began to have greater success in holding it in its designated place. As I picked up the pace of the stroke, we began to make headway against the tide.

"Now, stand up and do the same thing, but keep control of your stroke and let the water help you." I did as he said, spreading my feet for balance and beginning to make bolder and more controlled strokes. The skiff responded as never before, and in minutes we were up to the dock. Without another word, Jack jumped up onto the *Olive* and then onto the dock. Next thing I knew, he was walking back to his shack without so much as a backward glance. I let go of the *Olive* and drifted back into the tide to practice my new skill.

A couple of weeks later, after a very long and very hot day at sea hauling traps off Narragansett, as well as two traps on the inside and the offshore sides of the east wall, bailing tons of fish, we finally took a break for our noon meal. Cap gave us an extra half-hour and, instead of heading into Grandma's kitchen, Jack said rather gruffly, "Come with me" and headed toward his shack. I began to fall into his stride, which was not too difficult, as I was now nearly full grown (at least I thought so), and was close to Jack's full height. He was about five-six, with a strong, stocky build.

Jack lived in a strange-looking building that had the appearance of an old wheelhouse from a trawler. It had a couple of add-on rooms that jutted out this way and that. I had never been inside this house, and to the best of my knowledge the only other person who had was Cap—and that rarely. As we neared Jack's home I became quite inquisitive as to what I would find inside. My youthful imagination ran the full range from pirate treasure in a barnacle-encrusted sea chest, to dark, scary, dungeon-like rooms. Neither was forthcoming. As we climbed a short ramp to a deck that went all around the building, I became aware of some potted plants that would be the envy of any sea captain's wife. Opening the solidly built door, strong enough to withstand most gales, we entered a main room that had a low ceiling, with heavy exposed beams crossing from one side to the other. On the eastern end of the building there was a series of windows overlooking Salt Pond. You could watch the boat traffic from High Point to the north right down to Galilee and the breachway leading out to the seawall.

The interior, although festooned with a variety of nautical implements—a compass, a ship's wheel, a table with a few charts rolled out, parallel rules and dividers—was quite bright and downright cheery. I was amazed. I had no idea that Jack lived in or even enjoyed such clean, comfortable quarters. His dour appearance would not have led you to such a conclusion.

Without a word, he snapped a clean tablecloth onto the ruggedly built table in the center of the room and quickly set plates, cups, and silver for both of us. He stoked the coal fire under a large cast-iron pot sitting on the stove, then took a kettle off and poured water into a pan, inviting me to wash up before the meal.

Jack followed suit; then he bailed some hot stew from the iron pot into some heavy china bowls, made some tea, and began to eat—all of this with hardly a word.

When the meal was finished Jack cleared the table, put the dishes into the black slate sink, pumped some cold water onto them, and said, "We'll let them soak a while." He walked over to what appeared to be his chair, as it was well worn and angled so he could look out a window that covered a view of the dock, Cap's house, the huge "Ice House" barn, and a good part of the beach where we hauled our nets out to dry and repair.

I enjoyed the meal and appreciated the invitation, but I was still a little confused as to why I was there. Still, I knew better than to begin asking questions. Jack harrumphed a couple of times and invited me to sit down.

"What do you think of my boat?" he growled softly. Jack had an eighteen-foot open-cockpit boat with an eight-foot beam. She drew about four and a half feet—a deep boat for her size. She had the prettiest lines I had ever seen on a cockpit boat, a beautiful sheer and a bow with flare that would lift her as she was going into a sea and throw any white water away, keeping those inboard dry.

"She's beautiful," I stammered.

"She's yours then," Jack said. I couldn't believe him. "Of course I'll have to run it by Cap, but I'm sure it'll be okay."

I was dumbfounded and grateful, but afraid to ask why Jack was going to do this. He could see that I was about to explode, so he began to speak. He said more on this occasion than at any other time since I'd known him."I'm not one for talking much—but I like you. I have since we began working together.

"You work like a man, you don't complain when things are going tough, and most of all, you don't jabber on like most kids your age. God! I can't stand that jabber, and the questions, those never-ending questions. You don't do that. You're quick to learn. You respect your grandparents and the men you work with. You're quiet. I like that in a man."

I just listened. I couldn't believe my ears, but I sat there not saying a thing. He liked that.

"I'm getting done with trapping. I'm giving Cap notice at the end of the season, which is shortly, and I'm feeling poorly. I can't do the work that needs to be done. I won't hang on anyplace if I can't do the work.

"We've talked enough. Let's get back to work."

He raised himself much more slowly than I had ever seen before. He checked the fire, pushed the stew to the back end of the stove, filled the teakettle, and headed for the door. I followed, and as we walked back toward the dock and our work, I began to ask, "Why?" But Jack raised his hand, and I knew that he did not wish to discuss it any further.

When we got to the dock, we walked out to his boat, lying there moving gently as the flood tide began to lift her ever so easily at her mooring. "Take good care of her," he said. "She's always taken good care of me." Then he turned and went back to work.

Was Jack Lewis a lonesome man, a loner, or just a quiet man with a generous and giving heart who enjoyed company that didn't talk?

THE GREATCOAT ADVENTURE

My grandfather was a character with the best of them. One year the trap-fishing season had been very successful. All of the frames, barrels, and anchors had been pulled from their various positions along the shorelines that had been licensed to Cap by the state of Rhode Island. Some three months had been required to repair the twine bodies of the traps, to spread the repaired nets out in the fields to dry, and then to haul the massive nets up into the loft of the Ice House for storage until spring.

The history of the Ice House is interesting. When Cap Clark first arrived in Snug Harbor and began to apply his entrepreneurial skills to the fishing industry, there was no source of ice to keep fish fresh while it was being shipped to markets in Boston, Providence, and New York. The nearest ice plant was in Providence, some thirty miles away, and the most economical transportation was the railroad. While trucking was accessible, the cost of moving it over the road was prohibitive.

In those early years, the winters were very cold in Rhode Island—cold enough to freeze Salt Pond, a body of water some seven miles long, running from Galilee to Wakefield. A number of important estuaries emptied brackish water into the pond on every ebb tide, and it was refreshed with salt water on every flood tide. Salt Pond would freeze over in early January and hold fast until late February. So, mindful of the opportunity to make ice in the various ponds and small lakes nearby, Cap decided to take matters into his own hands. After receiving permission from the landowners abutting the bodies of water needed for this ambitious undertaking, Cap laid plans for the next segment of his project.

Knowing that he needed a huge building in which to hold all of his fish traps and their various accoutrements, as well as one large enough to keep ice frozen throughout the warm summer months, Cap designed the Ice House. The framing was twelve inches thick, deep enough to hold ample sawdust to insulate the building against the salt air and summer heat outside. The first floor was twelve feet high and insulated, as well. The second story was a massive, hip-style roof sitting upon sidewalls measuring ten feet in height. The end walls that ran up to the peaks on both ends were upwards of thirty feet high.

For the eastern end wall of the second story, Cap designed a huge insulated door that rolled open alongside the outside edge of the building. About eight feet above the large lower door, a second smaller door was installed, and just above that uppermost door, an imposing boom equipped with a large block was built into the structure. From the sheave of that block ran a wire cable that could haul both ice and nets up to the second and third floors for storage and/or removal, as needed. When completed, this structure was the largest commercial building along the coast from Stonington, Connecticut, to Newport, Rhode Island.

Cap had a special, well-trained crew that would go to the various ponds, cut blocks of ice weighing three hundred pounds each, and then transport them to the Ice House. They would hoist the ice to the second floor and stack it three blocks high, allowing room for sawdust to be poured in on top of the ice. Above that, the third floor held all of the netting and associated trap gear. On the ground floor, ice was stacked from the floor almost to the ceiling, leaving adequate space for additional poured-in insulation.

The design of this building was so efficient that an ample supply of ice was available into and throughout the summer fishing season. Frequently there was enough surplus that other fishermen could be supplied, as well.

As the years went along and ice-making machines came on the market, Cap set one up right on his docks and ceased cutting ice and storing it in his massive barn. But his trap fishing was so successful that his gear and equipment consumed the unused space, and there was ample room to mend nets under cover in the colder months.

Now, Cap, because his nature required him to be working at something, fitted out the *Olive*, his thirty-foot western-rigged boat, for dragging. He installed a compact winch that was nonetheless big enough to haul a set of small doors and a trawl from the ocean bottom to the deck of the boat. He did not want to hire a crewman to work with him, so he rigged everything close enough together that he could reach the ship's wheel, the engine throttle, and the winch controls from his position at the winch head. Thus, without moving much, he could set the net over the side, drop the drag doors, hook up the tow wires, and then reverse everything when it was time to haul in the net after a period of towing it along the seabed.

After fishing for a number of days and making small adjustments, Cap felt that he had a smooth-working boat, and he was confident that he would be able to work through the winter and on into early spring, when it would be time to get ready for trap fishing.

Because he was operating alone, and because the *Olive* was not built as ruggedly as the other small trawlers in Galilee, Cap decided that he would fish inside the Point Judith breakwater, which would provide him with complete protection from the winter winds and seas. Offering further shelter would be nearby landfalls to the north and west.

Cap set out one December morning from Snug Harbor and set a course toward the breachway between Galilee to the east and Jerusalem to the west. As he cleared the opening, he had to steer as close as possible to the west wall of the breakwater in order to get by the sandbar that nearly closed the channel at this time of the year. Once he was by the second buoy that was just inside the West Gap at the end of the wall, he turned and headed directly for the East Gap a couple of miles away.

Heading due east, Cap watched for some buildings, a water tower, and a huge oak tree that he used as range markers to show him the easternmost end of his tow. Aligning these inshore landmarks, as well as the Point Judith lighthouse and a beacon light on the eastern wall, he set his trawl and doors as he headed back to the west end of the Harbor of Refuge.

When he hauled his net in and winched the cod end (the very end of the funnel-shaped trawl, where the fish were concentrated) aboard, he saw several bushels of blackback flounders, a few pond eels, a couple of horseshoe crabs, and a half-dozen lobsters that were all keepers.

Then the wind began to freshen from the northwest, and with it came a bone-chilling cold, so before setting his net out for the second tow, Cap ducked into the wheelhouse to drink a quick cup of coffee from his thermos and to put on the army greatcoat that he brought home from Europe during the First World War. This was, as its name implies, huge in size, going from his ankles right up to and slightly beyond his ears. Made from the very best woolen fabric available, its weight was considerable, and it provided warmth even when soaking wet. The coat was a horrible brown in color, but Cap never gave a thought to that when he was shivering from the cold.

Warm once again, my grandfather set his net for the second time and again towed toward the west. As soon as the trawl was on the bottom, he put the becket, a rope lanyard, over the ship's wheel to hold the *Olive* on the predetermined course, then he stepped out of the wheelhouse onto the deck and began picking over the fish. He was pleased with the yield of the first tow, and as he iced down the catch in the sugar barrels he had secured on deck for just this purpose, he looked forward to an even bigger haul. He had set out this time with a flood tide and felt that the flounder would be coming up out of the sand to feed on an influx of bait brought in with the favorable flow of water.

As hoped for, the net was extremely heavy, and the running gear started to chatter with the strain as he hauled the doors up to the short boom hanging over the starboard side. Cap secured them, then started the winch to hoist the net up to the block overhead. As he was hauling back, it crossed his mind that his deck layout was working quite well; he was able to reach everything without straining.

Greatly aware of the potential danger of working alone on deck—especially when hauling in the net, what with all the rope dropping to the deck near his feet—Cap was mindful of his progress. Or so he thought. When he could see the cod end, he took a few moments to clear away any twine (netting) that might hinder the "bag" as it cleared the rail and was swung inboard.

Thinking that everything was shipshape, Cap went back to the winch and began to haul the cod end aboard. Removing the rope from the cleat, he took three turns from the block and falls, and wrapped them around the winch head. And as he slowly hauled the bag of fish upward, he would throw a coil of the rope behind him, making sure that he was keeping his feet from becoming entangled. The *Olive* began to list to starboard as the heavy cod end rose higher and higher. Knowing that the bag of fish would swing quickly inboard as soon as it cleared the rail, Cap prepared to throw off the turns of rope on the winch head as fast as possible;

this would allow the heavy cod end to drop onto the deck and prevent it from swinging clear of the opposite rail and out into space.

The *Olive* took a quick list as the bag of fish swung slightly away from the rail of the boat, causing Cap to step backward, unaware that he had put his foot into one of the rope coils, and as he did so, the bag came in over the rail. In a flash Cap flipped the rope off the winch head and let the bag drop quickly to the deck. Suddenly the coil of rope tightened around his leg, and in another flash it snatched him upside down and then heavenward, toward the block hanging overhead. Just as Cap's foot was about to be torn off his leg, the bag of fish settled on deck and went no farther.

During this maneuver, Cap's army greatcoat flopped downward toward the deck and engulfed his body; cutting off his vision and nearly his hearing as the heavy, wet wool acted as a baffle to the sound around him.

What a predicament! The bag of fish was rolling around on deck; Cap was swinging to and fro with the movement of the seas while hanging upside down in the rigging. Still in gear and controlled by the becket on the ship's wheel, the *Olive* turned ever so slowly in a circle. The tide was nudging the boat slowly toward the west wall, and the ground swell, as slight as it was, began to inch her toward the beach.

Of course, once the embarrassment of the moment had passed, the discomfort in my grandfather's left ankle began to overwhelm him, as did the painful spreading of his legs as he twisted slowly in the breeze and rolling seas. His concern for the safety of his boat evaporated as waves of pain took control. Time passed slowly—very, very slowly.

"Ahoy, Cap! Are you aboard?" yelled Joe Whaley on the *Virginia Marise* out of Galilee. At first, Joe did not see my grandfather hanging from the rigging as the army greatcoat enshrouded him from head to toe. Then Joe heard a muffled sound from overhead, looked up, and quickly realized what had happened. While Joe held his boat alongside the *Olive* he had his crewman, Carl Westcott, jump on board and throw the *Olive*'s engine out of gear. Then, he slowly lowered my grandfather to the deck. Cap emerged from the army greatcoat and painfully arose, leaning against the winch and mast. He took a quick look around and saw that he was rapidly drawing down onto the sandbar near the wall.

"Thanks, Joe. Thanks, Carl. I'll follow you into your dock as soon as I get this old girl heading in the right direction."

With that, he took the becket off the wheel, put the engine back in gear, and moved the throttle up to the "rusty notch" (the setting that gave him full speed). Cap spun the wheel to port just as the keel of the *Olive* rubbed ever so gently over the offshore side of the sandbar that he had avoided in the early-morning hours as he steamed out through the breachway. Rather than trip (open) the cod end and expose the catch to those he knew would be standing around Joe's place waiting to tease him about his little misadventure, my grandfather decided that he would wait until he got back to his own dock. Cap just eased the bow of his boat up to the side of Joe Whaley's dock, kept the *Olive* in gear and stuck her stem (the leading edge of the bow) behind a large piling. Then he walked out of the wheelhouse onto the deck and again thanked Joe and Carl for their able assistance.

"Well, J. E., I guess you'll be a little anxious to get back to trapping come spring, won't you?" said Joe, with a little grin on his face.

"Yes, Joe, I certainly will," answered my grandfather.

"Hey, Cap!" called Carl with an obvious chuckle in his voice. "If you ever want to sell that greatcoat, let me know. I just might want to buy it. The story that goes with it should be worth a few bucks, don't you think?"

"Yes, Carl, but I'll tell you one thing. I was embarrassed, slightly bruised, terribly confused, but I was never cold. So I think I'll keep the coat."

With the bantering coming to an end, old J. E. went into the wheelhouse, threw the engine out of gear, then backed the *Olive* away from the dock. In just a few minutes he was heading up the channel to Snug Harbor.

Other than wondering what kind of tow he had sitting in the cod end, Cap's primary thoughts were on preparing for another season of trap fishing. He was also wondering if he could keep this day's excitement under wraps, or whether he should own up to it among family and friends.

CAPTAIN HAMMY

Galilee was the home port of a number of fishermen who engaged in tub-trawling for cod and other groundfish during the winter, and in early spring of every year they hauled their tubs, hooks, lines, marker flags, anchors, dories, and various and sundry other equipment ashore. The goal of this activity was to clear the decks of their small, open-cockpit boats in preparation for the onslaught of the "city boys."

Most skippers detested the switchover to sportfishing, but there was usually more money to be made in a shorter period of time if they adapted to the whims of the landlubbers. They did so grudgingly, however, complaining at every turn. Not so Hammy. He was an exception to the rule.

About six feet tall and weighing in at two hundred pounds of muscle wound as tight as wire cable on a winch drum, Hammy had a face, hands, and body that had been weathered by wind, salt, and rough seas for fifty-plus years. No man or storm had beaten Hammy down or turned him from his chosen course in life. The soul within was of a happy and generous nature, quick to lend a helping hand to anyone in need. He had, over the seasons at sea, hauled more than a few mates out of stormy waters or quickly extended a towline to a boat that was in distress and was rolling around in a heavy ground swell. Hammy was liked and respected by all who knew him.

There was a modicum of envy, though, if the whole truth be known, as Hammy out-fished every other skipper in Galilee. He could set more tubs of baited hooks faster, haul them in with the fewest foul-ups, and retrieve more groundfish that any fisherman in the fleet. Hammy never boasted about these skills, though, and he never compared his catch with anyone else's.

He had been well seasoned in his lifetime of fighting the sea and everything produced by that loving, hateful, and unpredictable feminine force, and his passion for her strengthened each day that he faced her. He knew that if he ever ignored her, his very life would quickly be in jeopardy.

Hammy was renowned in Galilee as having an unusual sense for just where the best bottom fishing would be on any given day. The usual species of fish he and the others caught were blackback flounder, scup, sea bass, and tautog, but the goal of every landlubber with a line, sinker, and squid for bait was a codfish. Better yet, a steaker cod or the nirvana of them all, a whale cod.

Hammy was skipper of the *Galilee Striker,* and he was exceptionally popular with the summer visitors from Cranston and beyond because he provided them with the very best opportunity to fish the various grounds around Point Judith: Nebraska Shoal, the Whistle Buoy, the East Ground off Block Island, and as far offshore as the Deep Hole.

Those city boys would roll around on deck, puke over the rail every two minutes, and fight tangled lines all the day long in hopes of catching a whale cod. As the name indicates, this fish would run from fifty to sixty pounds while the usual, market-sized cod most often landed went about two to six pounds.

If you had a reputation for fulfilling dreams, you had all of the paying customers you ever needed. This was the happy situation in which Hammy found himself, trip after trip and season after season.

Once his boat was filled to capacity, the remaining body of long-faced landlubbers would drift down the bulkhead to other party boats, which would be waiting for Hammy's leftovers with their engines running. They weren't too proud to be second best or third or fourth.

One day a new party boat called the *Star of Portugal* appeared in the port, skippered by a captain totally unknown to any of the regulars in the local fleet. Although no one recognized him, newcomer Manuel Silva from Newport, Rhode Island knew each and every one of them. Several seasons earlier this stranger had set out to study each boat, each captain, and the fishing record of everyone in the harbor—especially Hammy. He had even palmed himself off as a city boy, hand line and all. He watched where Hammy went, when he went, and eventually even *why* he went to certain fishing grounds at certain times. He studied the tides as Hammy fished them, the reading of the barometer, the clouds, and the color of the water, then he matched that information with the catch brought ashore by everyone aboard. He made meticulous notes in his logbook.

Yet with all of the detailed information that Manuel had accrued over several seasons of fishing, he was still not able to figure out how to best Hammy. What could bring this giant of the Atlantic down? What was needed to unseat the highliner of Galilee?

One very important bit of information came to his attention in the most serendipitous way. While visiting a local watering hole, the Bon View, he overheard the bartender and one of the local fishermen laughing over the one superstition that held Hammy in unbounded fear. Manny felt the delightful taste of victory permeate his very being.

"I've won" he gloated as he slid off the barstool and headed back to his boat, anxious to put his devious plan into operation.

The very next Saturday, when the docks were crowded with landlubbers lining up to go aboard the *Galilee Striker,* and before Hammy got down to the boat, Manny struck. *Whack*! went his knife as the blade sank deep into the mast of Hammy's boat, pinning a note there. Manny then went ashore, boarded his own boat, and brought her right alongside Hammy's. He left his engine running; he had brought aboard ample bait, ice, and extra lines in case anyone lost his. Then he waited.

It wasn't long before Hammy came walking along the bulkhead, whistling and greeting many of his long-standing customers by name, cheerily saying what a great day it was going to be for bottom fishing.

With a big smile he said, "You'll all get a whale cod today. I feel it in my bones." As he said this, he came to his boat and, reaching up to grab a side stay, he gingerly swung himself aboard. Without hesitation he headed toward the companionway that led to the engine room.

Just then his eyes caught sight of something fluttering. He turned slightly to determine what was stuck to the mast. Suddenly he began to scream to everyone already aboard, "Get off the boat! Get off!" Then he called out to those about to come aboard, "Stop! Keep off! We ain't going nowhere today! Go away! Let me off the boat! Get out of my way! Get out of my way!" Hammy screamed these orders as he was running back up the bulkhead toward his shanty on the edge of the marsh, just inshore from the docks.

Manny then called out to those very disappointed city boys, "Come on aboard my boat today. I guarantee that you will catch as many fish as you would have with Hammy. If you don't, I'll return your money."

Nearly all of the confused crowd accepted his invitation and went aboard the *Star of Portugal.* With a smile of victory, Manny steamed out of the harbor, heading for Deep Hole.

What was it that Manny had used to frighten Hammy into such a state of panic? What was it that had struck terror in the heart of this stalwart seaman of fame who had faced the most powerful storms that Nature had thrown at him?

One of Hammy's longtime customers and friends jumped onto the deck of the *Galilee Striker* and went to the door leading to the engine room. As he approached he, too, saw what was stuck to the mast with a seaman's knife. It was a picture of a pig. That's all. Just a pig. Hammy had fallen victim to his irrational attitude toward (should I say it?) pigs, porkers, hogs, or—as he, with his limited vocabulary, described this much-feared creature, "those ugly things with the curly tails." That's how Hammy got his nickname. Of course, no one dared to call him that to his face.

Without question, Hammy was an outstanding example of mixing superstition and the sea. A few more ideas held by otherwise very intelligent individuals were equally interesting to me as a boy, and later, as the captain of my own vessel.

If the crew of a fishing boat observed a suitcase being brought aboard, it immediately went over the side, possibly along with the person who had caused this infraction of "the rules of the sea."

No crewman should be found climbing into any pajamas—silk, cotton, or any other fabric; he would probably beat the suitcase into the water.

There was no question in the mind of any fisherman that if the cook made a beef stew on board, there would be a Force 10 wind within twenty-four hours. This dish was commonly called a "Windjammer," and—as indicated—the name goes back to the days of sail. But I love beef stew, so how did I get around this obstacle? I had my wife make it at home. That doesn't count, I guess, because it never brought on bad weather—at least, not so's you'd notice.

No mast could be stepped (set into the keel or into a metal boot on deck) without a silver coin, preferably a silver dollar, being placed under it before the spar was lowered into place. I was replacing a mast on my second vessel, the *Dorothy & Betty II,* when I heard a man screaming at the top of his lungs and running toward the boat, waving his arms wildly in the air. I thought someone had died, or was about to. The crane operator stopped the process of lowering the mast for a moment, allowing just enough time for this guy—who turned out to be the harbormaster—to jump on board and throw a coin into the slurry of tar and oil previously poured into the boot as a preservative. Down into the goo the mast went, and at the costly rate of the crane, I didn't stop to pull the coin out.

Red, black, or green; big or small—no matter what its color or size—no umbrella was allowed on board a fishing boat.

Hatch covers were usually built in sections just large enough for two men to handle. When you removed the covers, you stacked them just the way they came off the fish hold. You never, *ever* turned them over. If you did, you would bring down the wrath of Neptune. At the very least, you would frighten the whole crew—except for the gang aboard the *Mary Alice*, owned and captained by Fred Gamache. His hatch covers were hinged, and the only way you got down into the fish hold was . . . that's right . . . by turning over the two halves of the hatch cover. Nothing really bad ever happened aboard Fred's boat, yet that didn't change the minds of the crewmen aboard every other vessel.

Also anathema to any seafarer was whistling while at sea, and—by the way—this was the only superstition that I could make any sense of. Years ago, there were only a few electronic navigational devices, but most all fishing boats had radio direction finders (RDFs) on board. In a storm or fog, when lighthouse beacons and buoy lights could not be seen, you could pick up and hone in on one or more shore-based radio signals, each of which sounded like a low, monotonal whistle. As you turned the small antenna on top of the RDF, it would indicate your bearing relative to the source of the signal.

The sound would get louder as you veered away from the signal heading, and grow softer the closer you got to it. If someone was whistling in the wheelhouse while the captain was trying to locate the proper course to the navigation aid being sought, he could easily get confused as to his location and possibly cause a wreck. Such a disaster was not likely, but it was possible. So we never whistled on board. Ever. Now, more sophisticated electronic equipment has replaced the old radio direction finder. In fact, I doubt that any of the fishermen sailing today would even know what one was.

THE STORYTELLER

"New-fun-land." That's how Johnny Long pronounced the name of his homeland, and he always said it with great pride. I never knew why he didn't go back home when he reached retirement age, although he never really retired. To the best of my recollection, he worked right up to the day he died.

He and my grandfather were old shipmates and close friends, and they continued to be long after I left Snug Harbor. A local fisherman, John Sisson, took a picture of my grandfather and Johnny working together, repairing a fish trap, and at a later date, he did an oil painting from that photo. How well it captured the very character of both men.

Both were craggy in appearance, with deeply tanned faces and hands, which in turn were gnarled and misshapen from arthritis and years of hard work. Yet the pleasure they found in working together at something they both loved to do is reflected clearly in the painting. I think that John Sisson's being a commercial fisherman himself allowed him to capture the very essence of the two "old salts."

Johnny Long, right (shown here with fellow draggerman Thad Holburton), hailed from "New-fun-land," as he pronounced it, and was one of the best storytellers the author ever heard.

Johnny Long lived in a shack about twenty feet square. It eased out into a freshwater bog, and the back end of this house sat on several pilings, like so many shanties in that area. Inside, it was just one large room, with two windows in each wall, allowing the sunlight to penetrate the cigarette and wood smoke that hung from floor to ceiling all the time. Even when you opened the one door in the place, any fresh air that tried to nudge its way into the room was stopped at the threshold.

The one exception was when we got a blow from the nor'east. That was the direction the front of the shanty faced and a good thing at that, as it gave a slight respite to anyone trying to wrench some oxygen out of the noxious fumes. In one corner of the room, Johnny had a grimy, unkempt bed that served as a luxurious place of rest for the biggest, shaggiest, dirtiest, and friendliest sheepdog I have ever seen. He was Johnny's loyal and constant companion. You never saw John without him, and if the dog was not welcome, then neither was John, and he pretty well told you so.

In the center of the room was a round table that took up a large portion of the remaining space and was the hub of activity in John's life. Those stopping by his shanty were invited to sit on any uncluttered seat, if one could be found. If a particular chair was cluttered, you just tipped it until it became uncluttered, then you slowly pushed the fallen items away in one direction or another—it didn't matter where. It was on that table that visitors would set down a bag of beer bottles (there were no six-packs in those days) and any edible goodies they had brought; then out would come the playing cards.

Six or eight fishermen would gather around the table and deal a hand of poker. Some would play cards and some would just drink beer and talk. Everyone would try to outdo everyone else in storytelling, and that's when I would grab a swordfish keg and set it down as close to the men as I could get. I would sit hunched over, put my arms on the table, and listen to every word that was spoken. All of the men were good at spinning yarns, but no one could match the man from New-fun-land.

When Johnny Long began a story, the telling took all of his concentration, so he would quickly put down the cards—even if he had a good hand to play. He would pull his chair out to the middle of the room, scrunch down even smaller than his already diminutive stature, and pull his grimy tam-o'-shanter forward a bit. I think he did that so you couldn't look directly into his eyes to discern whether he was telling the truth or a real whopper. He would take a long slug of his beer, wipe his mouth with the end of his dirty sleeve, clear his throat, and begin his story. Everyone, even those who had heard the tale many times, would become quiet because a real drama was about to unfold before their very eyes. All of us knew we were in the presence of a master storyteller.

In one of these tales, John was but a lad, working in the galley of a dory trawler as a cook's helper. On this particular trip to the Grand Banks, the crew had found the fishing really "sparse," and they had to spend so much time on the fishing grounds that they were nearly out of food. Meals were limited to potatoes and salt cod. After a few weeks of this very restricted menu, the crew members were beginning to get ugly, and since John was the youngest and smallest, he took the brunt of the complaints and roughhousing.

One very early morning, after a breakfast of potatoes and salt cod, the crew launched their dories and rowed out in different directions from the "mother ship" in hopes of striking a school of cod big enough to fill the schooner's empty fish hold. Johnny helped the cook clean up in the fo'c'sle and prepare for the noon meal—more salt cod and potatoes. When he had finished with his chores, Johnny had a few spare minutes on deck before he was to begin baiting some tubs of trawl for the afternoon fishing.

He happened to look to the west'ard and thought he saw something unusual on the horizon, so he asked the skipper if he could use the spy-glass. Given permission, John found the object just a little closer, and it looked like a large tree or bush of some kind. John asked the captain if he could lower a dory and row out to see just what kind of tree this was. The captain, understanding the curiosity common to young lads, authorized him to do so.

John swung the dory overboard, lowered himself into it, and cast off. He began rowing toward the object, and noticed as he got a little closer that the tree had a bluish cast to it. He just couldn't figure what it was that he was looking at. He kept rowing and in about a half-hour, he drew alongside. To his amazement, he discovered that the mysterious "tree" was a huge blueberry bush, washed into the sea from New-fun-land; and it was loaded with the biggest, ripest blueberries that John had ever seen. He leaned over very carefully, so as not to tip the dory, and began to pull the berries off in great handfuls.

He picked and picked and filled the dory up to the gunwales—or, as John said, "gunnels"—until he was sure that if he put another blueberry into the boat it would sink. As he reluctantly left his marvelous and miraculous find, John could still see enough berries on the bush to fill a dozen dories, but he had to let it go in order to row back to the schooner with his "catch." Both the captain and the cook were flabbergasted to see Johnny's haul, and they happily helped him offload the berries.

When the crew came back to the schooner for the noon meal, they were still grumpy about breakfast and about eating the same thing for lunch. To their delight, the cook presented them with unlimited numbers of blueberry pancakes with blueberry syrup, blueberry muffins, and blueberry pies for dessert.

When told of Johnny's find and of his hard work in collecting the berries, the crew gave him a happy cheer and appreciative thanks.

John then told the captain that if the schooner overtook the huge blueberry bush and its still-ample burden of berries, using it as a marker, he was sure they would have better fishing. By this time, everyone believed everything that John said. So the crew hauled anchor, hoisted sail, and set off after the blueberry bush with the greatest of haste. One of the men in the topmast called out that the bush was close by, so all hands lowered their dories, loaded them with freshly baited tub trawls, and circled the bush, ever widening the circle. No sooner had the hooks gone to bottom than the men began to haul in codfish of every size, some of them whale cod weighing up to fifty or sixty pounds each.

At night the crew would tie a line to the wonderful bush from New-fun-land, so they wouldn't lose sight of it. Each day their success increased, and each day the berries fell into the sea. At the end of the third day, the last berry fell from the bush and the codfish disappeared, but by that time, the crew had caught, dressed out, and salted enough cod to fill the hold right up to the hatches. The captain had Johnny untie the line from the blueberry bush, and it slowly drifted away into the fogbank that was making up to the sou'west. The crew brought all of the dories aboard the schooner, secured them (and everything else on deck) for foul weather, then hauled anchor, raised all the sails, and set a course nor'west, a quarter west, toward their home port of St. John's.

The spirit of all aboard was joyful indeed: they had a hold full of top-quality fish, no hands had been lost, and there were enough blueberries left to supplement their normal staple of potatoes and salt cod during the long return trip. Not a complaint was heard in the fo'c'sle or on deck, and all was well in the world as the men thought of home, wives, and wages.

John turned into his bunk that last night on the Grand Banks happy that he had earned the respect of the captain, the cook, and the crew. As he drifted off to sleep, his thoughts were of the miracle blueberry bush. John knew that only New-fun-land could produce such a wonderful thing, and only a New-fun-lander like himself could have found it. Out of respect for that bush and the positive way it had affected his life, Johnny Long never ate another blueberry as long as he lived.

Having told us this story, Johnny would sit back in his chair, look right at me, and say, "No sir, I never did eat another blueberry, but I tell you this—I'll have another beer."

SINK OR SWIM

Upon reflection, I find it difficult to understand—let alone explain to anyone else—why I never learned to swim. I was on or around the ocean and its estuaries from earliest childhood. I was never afraid of the water. Later, I was in it and, at times, under it while on boats of various sizes, and I fell overboard more times than I can count. I just never could get the hang of swimming, as much as I once wanted to.

I remember well the effort put forth by my Uncle John to teach me. He worked at times for my grandfather, and he lived in the crew's quarters in the back end of Cap's house. Whenever he had a couple of free hours he would quickly don a swimming suit and run out to the end of the dock. He would climb up and onto one of the pilings, raise his arms high over his head, and dive cleanly into the water below, leaving nary a ripple. I watched and yearned to follow suit. Frequently, Uncle John would ask me if I wanted to learn how to swim; then he would help me climb onto his back. I would wrap my arms tightly around his neck as he dove into the water while I clung to him like a barnacle to a lobster.

Upon coming to the surface I would choke and sputter and spout seawater like a beached whale. John would then try to help me make the proper strokes—always without success. He would finally tire and drag me up to the ladder so that I could climb up onto the dock.

Several other members of my family tried unsuccessfully to teach me, until I would just refuse to learn. My grandfather had never found it necessary to learn how to swim, and he had sailed around the world many times without any problem. So I figured that I could do so as well. Finally, they all gave up on convincing me, much to my delight.

I remember one summer day when I was working with Sally Accarullo. Sal and his father were fish buyers from New Haven, Connecticut. They would take most, if not all, of the fish caught in my grandfather's traps. As we unloaded the work skiffs and the *Olive*, Sal would ice down the fish in barrels or boxes, cover them, and then use a hand truck to move the barrels directly into one of his trucks.

Sal had just loaded a barrel onto the hand truck when a couple of his men threw a box of fish on top of his load. When it struck, the impact threw Sal off balance, forcing him to step back quickly. As he did so, he bumped into me as I was bending over to untie one of the work skiffs. I went flying off the dock, just missing the skiff, which was a dozen or more feet below me. I landed with the grace of a gooney bird and disappeared below the surface. The tide was ebbing, and I felt it beginning to pull me out into deeper water. I held my breath, leaned into the tide, and tried to reach the dock pilings that I knew would hold me in place. I was finally able to grab one, whereupon I reached out for the next piling, then the next.

Neither Sal nor his workers could swim, and I was too far below the surface to be reached with a rope. He said later that they had watched with fascination as I worked my way up into shallow water by grabbing one piling after another without coming up for air. When I got near the surface, the men jumped into the waist-deep water to help me the rest of the way to shore.

Later, when my grandfather learned about my experience, he was pleased that I'd made it ashore but didn't seem very excited about the event. Then again, I can't ever remember his getting excited about much of anything.

THE FORMAL SITTING ROOM

My grandfather's house, sitting prominently behind a barrier of huge riprap walls on three sides, was, up to that point in my life, the largest that I had ever seen. It was two stories high with an attic that had not seen a visitor since the house was built.

Whenever I was allowed on the second floor, which was only under adult supervision, it looked like some huge castle or museum seen only in movies. The floors were made of oak and glistened with a waxy shine. All of the curtains were of the finest Irish linen from my grandmother's birthplace, County Westmeath, Ireland. The windows were higher than a man's head and half a fathom wide, and they glittered in the bright sunshine as it traversed from east to west, bringing in the clarity of reflected light from Salt Pond, which ebbed and flowed around Cap's house. There was a wide hallway that ran right down the middle of the house, from the entryway of the second-story porch on the east end to the entrance of the "formal sitting room" due west. French doors opened at either end of that hall, allowing sunlight to flow throughout the second floor; and with those waxed floors, it was a great place to run and slide in your stocking feet if the adults were out of sight.

The most outstanding feature of the second floor was the formal sitting room on the west end of the house. It was extremely large and had floor-to-ceiling windows on the south, west, and north sides. It was sunlit throughout the day, and when all of the lights were on in the evening, it shone like a beacon throughout Snug Harbor.

This was my grandmother's sanctuary. No wet boots laden with fish scales or clothing soaked in squid ink or needles filled with mending twine or knives with which to cut that twine were allowed above the first step of the second floor. No one, including Cap, was allowed to enter that most sacred place if he even smelled of fish, let alone carried the accoutrements of the fisheries.

My grandmother had to feed a crew of more than thirty men at breakfast time, and again at lunch. She had to provide snacks and coffee at mid-morning, as well, and it was not unusual for my grandfather to invite a fish dealer, a new captain in the area, or just a wandering tourist to share a meal at suppertime. Along with her oversight of the dockside restaurant, where she oftentimes prepared the major portion of the food served, these chores more than justified her own private place, and Cap never violated the brief time she enjoyed on the second floor. Nor would he allow anyone else to do so.

But as the years accumulated and with the changes brought about by age and a few extra pounds, it became more and more difficult for my grandmother to climb the stairs to reach her private space and her much-needed solitude. Finally she reluctantly relinquished her haven of peace, and she told my grandfather to rent out the second floor. She added only one caveat: "No fishermen."

Cap put an advertisement in the *Narragansett Times* describing the "apartment," and the very day the ad came out, a man named Sibley Smith and his wife stopped by to look at it. They climbed the stairs and, as they turned left at the top, they looked directly into the sitting room and were nearly blinded by the sunlight. They said that they would take the apartment without looking at the rest of the rooms. They were told how many rooms there were and what the rent would be, but those things were of little interest. It was that sitting room that had them so excited about moving in as soon as Cap would let them.

With deference to my grandmother's firm instructions about "fishermen," Cap tentatively asked what line of work Mr. Smith was in. "I'm an artist," said Smith, "and my wife is my model." Neither statement made an impression on my grandfather. He was not much interested in any profession but the fisheries, and that evening around the supper table he took the first opportunity to tell us (and, in particular, my grandmother) that he had rented "upstairs," and that the man did not fish. No further questions were asked.

Before the week was over, the artist and his family arrived: bags, baggage, paints, pallets, brushes, a young boy, and a Great Dane that was big enough to ride but was as gentle as could be. This entourage caused my grandmother some concern, but knowing that her husband had committed the apartment to the Smiths—and knowing how strongly my grandfather felt about giving his word—she said nothing, simply welcoming them to their new home.

Mr. Smith was a rather tall, slightly built man with brown hair and a matching mustache, a significant limp, and—to my ear—a delightful English accent. He called to my adventurous mind the flight commander of a British Spitfire squadron. His wife was a tall, lean, and very beautiful woman with a rather close-cropped cut of hair that was very fashionable in movies at that time.

In short order Mr. Smith set up his "studio" at the west end of the house, in the formal sitting room so loved by my grandmother.

The family was most pleasant and relatively quiet, except for the galloping about by the Great Dane. Every time Grammy heard the dog running around upstairs she could envision his toenails digging into that highly polished floor, and she would wince and mutter something in Gaelic.

A few days after the artist and his family moved in, toward the end of a long summer afternoon, I noticed a number of pickups in our driveway. That in and of itself was not unusual, as many fishermen parked their vehicles in our driveway (which was actually wide enough to be a road) when they were planning to work on their boats or going fishing. But this was different. Once the cars and trucks were parked, the men just hung over the fenders or sat on the tailgates. They were smoking and talking and not moving down to the dock and their boats.

I was too curious to let this go by, so I sauntered over toward the group to see just what was going on. As I was about to ask, I noticed that all of them, to a man, had their heads lifted up and were looking up at the second floor of my grandparents' house. I turned to see what was capturing their attention, and as I did so, the sun setting in the west flashed a brilliant beam of light onto Mrs. Smith. She was seated high off the floor on some kind of chair, draped over its back, and she was stark naked. All the while Mr. Smith was painting up a storm, trying to capture the fading sun as it cascaded over the nude body of that beautiful woman. The Smiths seemed totally oblivious of the gathering audience of very appreciative fishermen.

About that time, as I was becoming more aware of the beauties of life, my grandmother looked out the window of the kitchen. Curious, herself, about the group of silent men, she stepped out to see what was going on. She, too, followed their steady gaze until her eyes took in what was taking place in her formal sitting room.

Let it be known that my grandmother dispersed the group of "art lovers" in short order; she sent me into the house and climbed with painful yet quick steps up to the second floor, where she told Mr. Smith to either move his naked wife into the bedroom or put up some drapes over those lovely, floor-to-ceiling windows that were so appreciated by the artist.

There was no joy among the fishermen of Snug Harbor that night, and no display of Sibley Smith's talent or model thereafter.

CAPTAIN FISH-FISH

There was another interesting character who tied up at the dock of Gus Damascus and, like Manny Silva, he hailed from Newport. I know nothing of this man's family or of his own history. He, like so many of the fishermen who had an influence on my youth, was unusual in appearance and manner, but retained the most colorful of personalities and was a delight to observe.

He was named Fish-Fish. He was without doubt legally blind and wore glasses that were as thick as the bottoms of the original Coke bottles of years gone by. An observer watching Fish-Fish working on the dock as he strung bait and loaded lobster pots or climbed into or out of his boat, the *Lady,* would have had no idea of the disability suffered by this man. But once he left the security of a memorized pattern of work, he was as blind as a bat—not a normal, shortsighted bat in bright daylight, but a totally blind bat looking into the landing lights of an Air Lingus 707 coming into Shannon airport at midnight.

This man simply could not see. To help you grasp the difficulties that this condition caused, let's for a moment consider what transpired aboard the *Lady* from the time that Fish-Fish threw off the lines and aimed them for the dock, totally missing the intended target.

The boat would quickly be caught up in the heavy current flowing by in that area. If the tide were ebbing, Fish-Fish would exit the breachway and continue on precariously along the west wall of the breakwater. Then he would head out to sea and begin to search for his pots.

In the event that the tide was flooding, then time and circumstance would sweep Captain Fish-Fish northward, toward Snug Harbor. Whenever another fisherman, ashore or afloat, saw Fish-Fish going by, he would run to the nearest vessel, steam out to the *Lady*, and—after determining where Fish-Fish wanted to be going, would get him turned around and heading in the general direction of his intended course.

One interesting idiosyncrasy with Fish-Fish was his ability to steer by compass. If he was in open water, he could find his way along a certain course with a modicum of accuracy. He would run his time on this course, keeping track by using an alarm clock (with a magnifying glass, he could just make out the already large numbers). When the alarm went off, he would take a wild guess that he was within the vicinity of his lobster pots. The color code used by Fish-Fish on his buoys was a glossy fire-engine red with a bright yellow band four inches from the butt end. Attached to each buoy was a two-foot-long, three-inch-wide, flat stick with the same colors. Most lobstermen working out of the Point used such an addition only if they were lobstering in an area of heavy tides, but Fish-Fish needed every advantage he could get.

In the event that he could find his pots, which seldom happened, he would begin the process of hauling them. Most often, though, he would circle the area for a brief period of time, then he would shut his engine down and listen for the sound of a Point-based day boat that would be dragging the Sou'west Ground or Charlestown or wherever. If he heard an engine running, he would then start his and head in the direction of the sound.

I've lost count of the times that we on the *Kathy Dick* would watch Fish-Fish approach us and then yell out in his most unusual voice, "You see my pots?" If we had, we would send him in the general direction of our sighting. He would steam off just as happy as a clam at high tide. I have no idea how successful Fish-Fish was at generating an income at this work, but he would return year after year to Jerusalem and spend another summer hunting down his pots.

He had one very successful escapade in his lobstering career during the later part of World War II, when he had put his very life in danger by setting a few trawls— several pots tied to a single line—on the west side of Block Island. The islanders were notorious for wreaking havoc on anyone who infringed on their territorial waters—net fishermen or lobstermen—but somehow he survived with life, limb, and fishing gear intact. The locals must have observed how very limited he was, and they were moved to overlook his transgressions.

At any rate, on this warm, clear July morning the sun had barely cleared the horizon, and Fish-Fish was wandering like a lost puppy looking for his pots with no one to assist him. There must have been a limit to the islanders' generosity. Although suffering from a severe dimming of his vision, he had a mental acuity that amazed people at times—and this was one of those times.

Fish-Fish struck a partially submerged object in deep water about halfway between Block Island and Montauk Point. He looked over the side of his boat and quickly recognized the object as a periscope. His next action saved his life and perhaps untold numbers of others.

He continued to haul and reset his pots as though he had seen nothing unusual. Steaming this way and that for the next hour or so, he then headed for Galilee.

Fish-Fish had no radio aboard his boat, so he had to get ashore as quickly as possible without creating concern aboard the submarine, which almost certainly belonged to the enemy. He was apparently successful in his charade.

Upon arrival at the Coast Guard Station, Fish-Fish told his story to the chief petty officer on duty. Now, this sailor was in a quandary: He knew Fish-Fish and the lobsterman's inability to tell right from left, and north from south. Should he now pass on information that he himself sincerely doubted? Well, he had been in the service long enough to know that you pass on what you don't understand and that you salute when the brass tells you to solve the problem. At any rate, Quonset Point Naval Air Station was put on alert. A convoy was due to come up through Long Island and pass the coasts of Connecticut and Rhode Island around noon of that very day. This sub, if there was one, was in a perfect position to attack the convoy.

The navy quickly dispatched bombers to the approximate location given by Fish-Fish, and they found it. The submarine, an Italian of all things, was severely damaged in the subsequent attack but was able to retreat to deeper water offshore. Based on the observations of the U.S. warplanes, it was suspected that the sub eventually sank.

At any rate, Fish-Fish was a local hero; he had kept many lives and ships from disaster. Very humbly, he spent the rest of that day . . . looking for his pots.

MEETING APHRODITE
(and other interesting characters)

I first learned of Aphrodite, the Greek goddess of love and beauty, in my study of ancient religions at LaSalle Academy. I had no idea that she really lived in my neighborhood.

Gus Damascus was a lobster fisherman from Newport, Rhode Island, who fished out of Jerusalem every summer. He had a dock that ran out into the channel that led into Salt Pond, and it was here that he secured his twenty-five-foot, open-cockpit boat *Athena*.

Since Gus fished the area around Point Judith and along the eastern shore of Narragansett, he found it more convenient to fish out of Salt Pond. He was also in need of lobster bait every day, so he had arranged for John Dykstra on the *Kathy Dick* to drop off a couple of "mackerel barrels" full of skates, brills (a small flatfish), and yellow eels. Gus and his stern man (deckhand) would string these up, then salt them down in the bait box on the deck of the *Athena*.

I was fifteen this particular summer, and Johnny D. had offered me a site on the *Kathy Dick* for the summer months. I jumped at the opportunity. My grandfather felt that the heyday of trap fishing had passed and in the years ahead would only fade as a practical means of earning any appreciable income. Therefore, this would be the first summer that I went dragging rather than trap fishing with my grandfather, Cap Clark, out of Snug Harbor.

Because John was so particular about how we handled the fish that we caught, there was always a ready market for our catch. Unlike most of the other day boats, we had a small fish hold in the *Kathy Dick;* when I stood in it, the upper edge would be chest-high. In it we carried crushed ice with which to ice down our fish as it came aboard. We put the catch into small side pens, covering them with ice, and we kept the lobster bait in a separate pen.

Working with Johnny D., I had on many occasions met and talked with Gus Damascus. He was as handsome a man as I had ever met. About five feet, eleven inches in height, he was brawny, with broad shoulders and arms that were rippling with well-used muscles, a result of hauling lobster pots and moving them around on deck all day. His voice was softer than you would expect from a seaman, with just a slight and pleasant Greek accent.

During the summer, Gus lived in a modest cottage just inshore from his dock. His family would spend part of the week in Newport and the balance of the time in Jerusalem with Gus. Because of their schedule I did not get to meet them, as we would stop by to deliver the lobster bait at different times during the late afternoon or early evening. We would usually drop off the barrels on Gus's dock and then head up to our berth in Snug Harbor.

On one especially lovely summer afternoon, the *Kathy Dick* stopped by the dock owned by Skip Streeter's fish market, a business that catered to the summer crowd visiting Jerusalem. This pier was two docks up from Gus's. As we pulled alongside, I was down in the fish hold getting ready to fill some fish boxes with flounder, fluke, sea bass, and a few scup, caught only hours before on the Sou'west Ground, less than an hour offshore.

My job was to straddle the box to be filled and then to drive a short-handled, two-tined pitchfork into the fish in the pen and drag or lift them into the box.

John would hand down a special unloading strap that I would put over both ends of the box. When this was done, John would slowly lift the box up out of the fish hold with the *Kathy Dick*'s boom and winch, then swing it over the side of the boat, whereupon one of Skip's men would catch hold of it and John would slowly lower it to the dock.

Just as I was about to fill up another box, an outstandingly beautiful, olive-skinned goddess stepped aboard the *Kathy Dick* and walked very carefully to the stern. She looked down at me in the fish hold and smiled. She then stepped up onto the stern bitt, balancing delicately on her toes; lifted her arms over her head (and firm, well-formed arms they were); and posed for what seemed like hours. She was wearing a one-piece bluish-green bathing suit—and wearing it very well, indeed. In my short life, I had never seen a more exciting woman.

As she balanced on that stern bitt, I just stood there with my mouth open and watched. At the moment when she lifted her body (that very lovely body) into the air as she dove, John hollered at me and I drove the pitchfork right into my leg. I just couldn't take my eyes off that vision of Greek mythology. I had seen Aphrodite, in the flesh, and very lovely flesh it was. She was the daughter of Gus Damascus. But, to my sorrow, I was never to see that bathing beauty again.

Fifty-five years have gone by, and I can see that remarkable image in my mind as clearly as on the day it appeared. And I still carry the scars to prove that the story is true.

Skip Streeter was another memorable character from my early days. The owner and operator of a local fish market, he sold wholesale as well as retail, and had built up a good business, although he himself did not catch all that he sold. He bought most of his product from my grandfather and the balance from Johnny D., or from a few of the hand-liners and lobster fishermen in the harbor

That was not always the case, though. Skip had been a partner with Harry Bannister, another fisherman from Jerusalem. Together, they had owned and operated a small (thirty-foot), western-rig dragger. A western rig is a boat with the wheelhouse forward, and I mention that because it's an important part of the story.

So is the fact that Skip stuttered—not just every so often, but with near constancy. This impairment became more noticeable by degrees as his nervousness increased.

Over the years, Skip had suffered embarrassment as local fishermen teased him about his problem, so he had developed a method to minimize his stutter. He would make a great effort to speak more slowly and would deliberately enunciate every word. He improved considerably over time, and the teasing eased somewhat. Just let him get uptight about something, however, and he would lose all control and would begin to spit and sputter to the point that no one could understand him. And when you are on a boat and in some tricky situation, everyone needs to understand you—clearly, definitively.

One occasion when clarity was very important in Skip's mind was the late-summer afternoon when he and Harry were steaming in on their boat, the *Sally B.*, from just sou'west of Nebraska Shoal, heading for the West Gap in the breakwater at Point Judith. Fog had engulfed them within minutes of their getting under way.

As Harry held the course at the wheel, running his time, Skip cleared the deck of fish and cleaned the net, getting it ready to be raised off the deck as soon as they reached the dock. Just as he was going to ice down the fish, the boat slowed and began to turn in a wide circle. "Skip, come in here for a minute," called Harry, with a tense tone in his voice. By now the fog was as "black as a whore's heart," as the saying goes along the docks. Skip was completely under control until he heard, and then saw, Harry in a sweat. "Wh-a-a-a-t's wr-o-o-ong, Ca-a-p?" Skip slowly enunciated.

"I need you to hold the wheel for me," Harry bristled, as he hated it when Skip responded this way. "Speak up, man, speak up!" Harry blatted out in growing frustration. "I missed the light at the end of the wall, and I turned her back on her wake. I'm just making the thirty-foot depth again, so I know we're just outside the wall. Come in here with me and watch the sounder as I edge her inshore. As soon as we make the breakers, I'll know just how far we have to steam before we come to the gap."

Reluctantly Skip complied, squeezing into the small area they boastfully called the wheelhouse. Tight places were another thing that made Skip nervous, and with Harry's bulk, the ship's wheel, the electronics, the compass, and the other equipment needed to run the boat, this was a *tight* place.

As soon as Skip took the helm, Harry decided to go on deck and work his way around to the forward side of the wheelhouse. He motioned to Skip to open the windows, so he could speak to his partner without raising his voice. The boat was making very little headway, but Skip could feel the transom begin to rise higher with each wave.

He could feel in his bones the nearness of the ocean bottom, and he could just see the shadows of the huge granite boulders that had been part of the original breakwater but had been torn away by hurricanes and gales over the many years since it was built.

"Skip, I think I can hear the waves on the beach," said Harry. "I'm going to grab the bow stay and lean out over the stem to catch sight of the

A young Sam Cottle (right) helps Captain John Dykstra haul a drag net (bottom trawl) over the rail of the Kathy Dick*—without the aid of a winch or even a block and tackle. Once the cod end of the net is emptied on deck, the author sorts through the catch.*

breakers before we get caught in them. Keep an eye on me, and be ready to turn as soon as you see me wave."

"W-e-ll, Ca-a-a-ptain, while you are a-a-a-alooking, I-I-I-I'm g-g-o-o-i-i-ng t-o-o t-t-u-u-u-r-rn h-e-er a-a-a-ro-u-u-und."

And that is just what Skip did—and none too soon, either. For just as the boat came about, one very large comber rolled over the *Sally B.*, washing Harry overboard and sweeping him right into the beach at Matunuck. Some men casting for striped bass in the heavy surf saw Harry thrashing in the water and hauled him out before he drowned.

Skip, on the other hand, was heading to the south'ard, and as he came abreast of the West Gap, the fog lifted just enough for him to make the turn around the end of the wall. Then it settled in again as heavy as before. But Skip was now confident that he could maneuver the *Sally B.* along the wall until he was safely inside the breachway and alongside his dock.

After this frightening experience, Skip sold his share of the boat to Harry, and he built and opened the fish market, never to go to sea again. In time he was also successful in controlling his stutter—mostly.

As characters are wont to do, they drift in and out of our life's experience and they enhance our joy in living. The foregoing individuals were among just a few of the threads in the tapestry of my life.

A STORM TO REMEMBER

September 4, 1938

The people of Cape Verde Island, off the west coast of Africa, were the first to observe the makings of the storm clouds. The water became gray and almost dappled in some areas. The local fishermen quickly became aware of the absence of the fish usually found in those waters at that time of the year.

September 4, 1938

In Rhode Island, Cap Clark was concerned with the lateness of the mackerel run as he and his crew hauled the last fish trap on the offshore bend of the middle wall, just south of the harbor of Galilee. He had been close to ending the trap season. With the shortage of fish moving along the shoreline and with the traps becoming heavy with seaweed and moss, he was about to haul all of his gear and call it quits for the year.

Then, however, finding a decent haul of mackerel and even a few hundred pounds of large scup in the net, Cap had a change of heart. He decided to change the "kitchen" and "parlor" sections of the trap because the twine was so laden with sea growth that the heavily corked headropes holding the nets above water were hardly showing.

This would mean a long day for the crew, as they would first have to unload the morning's haul at Cap's dock in Snug Harbor; then Cap would have to haggle with the various fish dealers over the highest price, fuel up the *Olive*, grab some grub, and head back to sea. Before making the turnaround, he and some men would load the trap skiffs with the brand-new twine to be used in the net. He would also need to bring out a couple of large trap barrels, as two of the existing ones were low in the water. They might have been struck by the *Olive* or perhaps by a party boat trying to fish inside the trap, hoping for a quick catch for the city boys on board.

As the *Olive* came alongside the large barrels on each corner of the "parlor," the crew quickly jumped into the skiffs assigned to them. Men in two skiffs started to cut the sections laden with seaweed free from the headrope of the trap. The rest of the crew, carrying the new twine, quickly began to lash it to the trap frame and to ease the clean netting overboard.

Dusk was beginning to settle when they finished the job. The men returned to the *Olive*, dug into their shirt pockets for packs of near-soggy cigarettes, fired them up, leaned back against the bulwarks, and relaxed for the first time since sunup.

Cap lit his Camel, hiked his booted foot up to the rise of the deck inside the wheelhouse, and while heading for home, felt contented in his decision to reset a couple of traps. He would set the other one on the east side of Point Jude light. This time of year, the bait that the mackerel fed on would hang over the shoal running nearly north and south, parallel to the peninsula leading out to the light on Point Judith.

The next day, Cap and his crew hauled the second trap that he wanted to have ready for the possible run of mackerel and scup he hoped would continue to move inshore. The smaller draggers, or day boats, often worked on mackerel for another month or two, so it seemed likely that Cap would catch his share if he were ready for them.

After the second trap had been set, Cap turned his attention to hauling the trap on the inside of the east wall, as well as the largest trap off Matunuck Beach. When the fish were bailed into the *Olive*, the trap frames were stripped of the heavy, seaweed-laden twine. With this done the crew could dry out the netting on the beach south of Strawberry Island.

Now that all but two of his traps had been hauled out of the ocean and stretched over the sandy beaches in and around Snug Harbor, Cap was anxious to get the twine dry, cleaned of the seaweed and moss, and put up in the Ice House, the big old barn behind his house. Until all the gear was stowed out of the weather, he would be nervous as a cat crossing Route One. The cotton twine, which was so very expensive, had to be completely clean of any residue from the sea or it would rot slowly all winter and would literally disintegrate when hauled out beneath the springtime sun.

"Yes sir," Cap said to himself, "I've got enough at risk with those two new traps out there when bad weather could make up any day. Well, this game is all risk anyhow, I guess."

September 10, 1938

The storm gradually increased in intensity as it dipped down toward the African coast and then ever so slowly turned out to sea.

September 10, 1938

The surface of the water inside the trap off the middle breakwater was quivering as the massive school of mackerel increased its speed, the fish circling and nudging every inch of the twine confining them.

As Cap called out some warnings to the crew about keeping the cork line hauled up tight to prevent any spillage of fish, he felt pretty good about his earlier decision to keep a couple of traps working.

"Well now," he thought, "if we can keep up this streak for a couple of weeks, it should offset that yacht's tearing up the gear at Matunuck. Yes sir, by gum, only two more weeks."

September 19, 1938

There are no weather satellites, no weather radar, no offshore weather buoys. Only shipping on the high seas keeping a weather eye on the storm and passing their observations on to the U.S. Weather Bureau.

The hurricane was tracked as it moved west from the African coast toward the Bahamas. The bureau knew it was powerful because it had reached Category 5 strength, but most every weatherman believed that this storm would curve back out to sea before reaching the Northeast.

As the hurricane approached the United States, an area of high pressure just east of the coast kept relatively close to land as it moved northeastward. One young research forecaster, Charlie Pierce, concluded that the storm would not continue to move northeast and head out to sea but would, instead, track due north. He was overruled by the more senior meteorologists, and the official forecast called for cloudy skies and gusty conditions—but no hurricane.

September 19, 1938

Cap Clark was the only one in the region who still had traps in the water and thankful he was. The last two weeks were "slammers." Both of the remaining traps were filled with mackerel and scup in what was an unusual run of fish this late in the season. With no other traps pouring their catches into Fulton Fish Market in New York, the price of fish was high and staying so.

There was talk about some wind and seas building offshore, but this did not concern Cap, as such weather conditions did not adversely affect his fishing, which took place right up on the beach. Normally when he heard of a tropical storm coming at this time of the year, he would be quick to haul in his gear, as well as running the *Olive* up Salt Pond to get behind some high ground on Great Island or into some other lee.

None of this was worrisome now, though. From all available reports, the bad weather was heading out to sea. Cap's primary concern was finding a few extra men to help in hauling traps this time of year. He had let his large crew go just about two weeks earlier, when he had thought that trapping was over for the season.

He doubted that the storm offshore would even affect the run of fish, but, ever cautious, Cap would watch the "glass," or barometer, to see what might develop in the days ahead. It was nearly noon and he was heading in with the day's catch. He was watching the load of mackerel as they sloshed around near the top of the deck checkers. Today's catch was smaller than those of the last few days, but it would be profitable enough.

Cap knew that he was pushing his luck some by keeping the traps in the water so late, and he planned to haul both sets in a couple of days. He still thought he had plenty of time.

September 21, 1938

What would soon be known as the Great New England Hurricane of 1938 was acting as predicted by Charlie Pierce: The storm did not continue to move northeast and then curve out to sea but did, indeed, track due north. Moreover, it accelerated in forward speed to seventy miles per hour, the fastest known in the history of hurricanes. Since hurricane winds rotate counterclockwise, those to the east of the eye are moving from south to north. Because this storm was moving in the same direction, its extraordinary forward speed added force to its already powerful winds. Eastern Long Island and Rhode Island were soon to be hit with wind speeds that would exceed 180 miles per hour.

Without warning, the hurricane struck Long Island around 3:30 P.M. just a few hours before high tide. Shortly thereafter, the storm wreaked its violence between Montauk Point, Long Island, and New Harbor, Block Island, reaching the breakwaters of the Harbor of Refuge at Point Judith, Rhode Island, within minutes thereafter.

September 21, 1938

In the early hours of this day Cap had his crew haul all of the work skiffs high up onto dry land. He secured all of the outbuildings, including the Ice House, which held all of the valuable trapping gear to be used next season. All surplus equipment still on the docks was covered with heavy tarps and tied down.

Cap instructed my grandmother to leave the main house, which was built upon a seawall just a few feet from Salt Pond and the dock area. He told her to get up to the summer cottage, which had been willed to my grandmother by Chef Wilbur, a former employer and special friend. The cottage was about a half-mile inland from the shore and sat on ground that was several hundred feet higher.

Contrary to Cap's instructions and strong warning that she should leave quickly, my grandmother decided to pack all of her Irish linen, the family pictures, and some favorite dishes and silver.

In the meantime, Cap and two hired hands fueled up the *Olive*, pulled the smallest work skiff up close to the stern, and struggled to haul aboard and place on the transom of the *Olive* one of the smallest of the trap anchors. They got out the life jackets; stowed some grub, coffee, and water; and then let all the lines go.

The pride of J. E. Clark swung away from the dock, heading up the channel and on into Salt Pond toward the safe anchorage behind the northeast side of Great Island. It had deep water and high, protective land to seaward. This would be the safest spot to weather the storm moving in from offshore.

Once in position, Cap had the crew take the ends of the heaviest lines available and row them ashore. Once on the island they climbed up the bank to the highest ground and tied one hawser around the trunk of a huge oak tree, well rooted in the earth. The second line was tied around a heavy, round boulder that had been used as a storm anchor many times before. They then returned to the *Olive* and, now that her bow was firmly secured, they slowly eased the trap anchor into the stern of the work skiff. When they moved it the proper distance from the stern of the *Olive*, they rolled the anchor overboard while holding firmly to the lines that would be tied to the larger boat's stern bitt. Then they hurried back to the *Olive*, hauled the skiff on board, secured it, tied the anchor down, and went below for a "mug-up," preparing to face the unknown.

My grandmother, who had been babysitting my cousin, Judy Mae, not yet two years old, continued to pack her most important possessions into her car, which she had turned around so it was heading up the driveway, toward higher and safer ground. She left the motor running during her final trip down to the car with the last of her belongings.

My grandmother had never experienced such high winds in all her days. She was not afraid until she noticed the water at the shoreline had receded somewhat, even though the tide was flooding. She did not understand just what that meant, but knew it was not natural, and was certainly not good. It was when my grandmother looked down toward Galilee that she saw the storm seas topping the seawall to a degree that she had never before experienced. She was standing with the baby in her arms, out on the second-floor porch, when she saw the largest wave she had ever witnessed crash into and over the seawall. Without any slowing, it rushed inshore toward the fishing villages of Galilee and Jerusalem. My grandmother screamed, hugged the baby close, and ran down the stairs to the car.

Quickly setting the child down on the front seat, she put the car into gear and floored it, racing toward higher ground. Just as she pulled into the summer-cottage driveway, my grandmother turned to look back at the home she had left only moments before. The giant wave of seawater, not losing speed or height, was crashing into the side of the house, tearing the porch on which she had just been standing right off the face of the building, carrying it, along with tons of other debris, right up Salt Pond.

What my grandmother had witnessed was just a representative sample of the power of this unexpected storm.

From Westerly, bordering Connecticut, to Sakonnet Point, on the eastern boundary of Narragansett Bay, the increasing wind and typhoonlike rain bombed the entire state of Rhode Island. The end of the day would find 312 men, women, and children dead or missing. One hundred million dollars' worth of damage had been wrought across this smallest state in the Union.

Not in recorded history had such a disaster visited these shores. The account of the storm can never emphasize too much the element or the degree of terror, pain, and personal loss resulting from this most fearful catastrophe ever to strike Rhode Island. If one were to identify a singular element that caused such a loss of lives it would have to be that everyone—the weather bureau, state government, weather-wise seamen—all were unaware of the ferocious storm that was cascading relentlessly toward the citizens of southern New England.

An example of this was a report from the weather bureau in Rhode Island at 3:00 P.M. on September 21 calling for winds of 40 to 45 miles per hour. Ten minutes later, a news service sent a telegraph to the *Providence Journal* warning of a hurricane with winds of 75 miles per hour or more. Very few citizens—let alone responsible authorities such as the state police—were aware of this upgrade in the warning.

Shortly after the first report of increasing winds, their velocity had increased to 87 miles per hour, and then to 121 miles per hour. The water being driven relentlessly before the storm was creating havoc in its forward movement. Many of those who had survived, even as family members were torn from their grasps, reported a tidal wave one hundred feet high. It was before the tidal-wave warning that a Rhode Island state trooper had been all around the summer-cottage area in Jerusalem, advising people to leave their homes and to go to higher ground. The way to safety was across the bridge at Potter Pond channel and then up to Route One, probably three miles or so away. The trooper headed toward that very bridge and the group of summer residents in the Potter Pond area, planning to advise them of the danger of the storm and to tell all of them to evacuate.

The trooper never had a chance to sound the warning. As he approached the bridge from the Jerusalem end he was driving in ever-higher water and blowing debris, and he was totally unaware that the bridge had been washed away. His car went over the edge and was quickly washed away in the tidal flood. The trooper's body was found several days later.

The storm careened across Galilee harbor, striking full force on the docks and buildings along the bulwark that had recently been built from the Coast Guard Station south. One family tried to head home to Great Island, less than a mile away. They had to cross a low causeway to get to the island, which was all safe, high ground. Their car stalled in the ever-deepening seawater, so they squeezed out through a window into the rising tide.

Frank Harrater took hold of his twelve-year-old son, Blakey, and screamed for his wife, Rachel, to hold the boy's other hand. The three of them began to plod through the chest-high water toward home—so near and yet so far—when a large timber swirling in the tide struck Rachel on the side, turned, and hit Blakey in the head, decapitating him. The boy's body was torn out of his parents' hands and was out of sight within moments in the rampaging torrent.

Several days later Blakey's parents, along with fishermen friends, were walking the mudflats on the east side of Great Island looking for the boy's body. "Niggy," Blakey's dog, was with the group when he found the boy's fingertips sticking out of the mud. Everyone started digging and, in moments, they found the body, crediting Niggy with the find.

Leo Joyle and his girlfriend were lost when the house they were in was ripped from its foundation and swept along in the torrent of seawater that had destroyed the huge sand dunes on the seaward side of Joyle's home. Neither the people nor the house were ever found.

No other deaths were reported in the village of Galilee, although some twenty-four fishing boats had been lost, and that made up nearly the entire fleet.

Nor were there any deaths in Snug Harbor, though the buildings on my grandfather's dock were either washed away or were badly stove up. The *Olive* and her crew survived a hair-raising adventure trying to keep the boat from busting free from her moorings and riding the incoming tide up Salt Pond to her destruction. They were successful, and the *Olive* survived to carry Cap Clark to new adventures.

The main house in which my grandparents lived lost the front porches on both first and second floors, but other than a few puncture wounds from drifting debris and a complete flooding of the first floor and all of its furnishings, the house survived to comfort, entertain, and protect family members and friends. She would be filled with laughter once more—as was fitting.

I was seven years old at the time of the '38 hurricane and was totally unaware of the seriousness of the event. To me it was a time of exciting things to see and do. For instance, Ken Knight had a grocery store just a little south of Cap's, and stocked a huge assortment of ice cream. The tidal wave pretty well devastated Ken's store and in doing so released great

The otherwise unnamed "Hurricane of '38" struck coastal Rhode Island with unexpected and devastating force, leaving deaths, vessel sinkings, upturned buildings, and other massive damage in its path. The Galilee fishing fleet was hit hard, losing twenty-four boats. The author, who was seven at the time, helped his grandparents rebuild the front porch of their house.

numbers of three-gallon tubs of chocolate, strawberry, coffee, and many other delightful flavors of ice cream.

Having been raised by seamen, I knew well the laws of salvage: what I found that had not been claimed by someone else was mine. So I found the biggest serving spoon my grandmother had in her silverware drawer and waded into battle with the tubs of ice cream that were scattered over the nearby backyards. I would find one container after another, open the cover, pour out any water that had seeped in, then begin eating. I knew that I could never eat everything in sight, so I took walloping big bites of one flavor before moving on to the next tub. Looking back on my actions now, I wonder how I failed to get some serious disease, as the liquid I poured out was a mixture of seawater, pond water, and likely a smidgen of residue from the local outhouses. At the time, none of these things crossed my mind—it was just fun.

One related story that was passed around after the storm involved the sight of Ken Knight, a stately man, and his frail little wife—I think her name was Jen—as they both waded through the rising water. Ken was carrying a large steel cash register. It was brand new and was full of cash from the day's sales. He was not about to lose either the money or the machine. Ken was striding through the water, making a wake as he moved, and poor Jen was holding onto Ken's wide Sam Browne belt for all she was worth, twisting all about in Ken's wake. Reportedly, he kept saying in a loud voice, "Hang on, Jen, we're almost there. Hang on!" They finally made the high ground, and Ken had three things of great value: the money, the cash register, and, of course, little Jen.

Very few humorous events happened on or closely following that day of September 21, 1938. What few there were brought needed relief to those who observed them, and were passed on to others in an effort to ease the pain felt by any who were old enough to be aware of the horrendous impact of this ravaging event.

On the day following the storm, one could step out of doors, look upward, and be graced by an azure sky marked by wispy white clouds drifting toward the horizon, both east and west. But when your eyes dropped to earth, the observer would encounter the appearance of a war-torn environment, with near-total destruction in every direction.

Cap Clark's fishing docks, which had run out to the very edge of the channel that passed up Salt Pond toward the town of Wakefield, were torn from the muddy bottom of the pond and were cast into the flood tide. Where they ended up is anyone's guess.

My grandmother's restaurant was resting on the seawall that Cap had built many years ago to protect the building from the ravages of the daily tides. The place was broken in half, with one end sitting on dry ground and the offshore end plunging into the changing tides. The fish market had been ripped completely off the dock and had plunged into the deep gully of water that had been carved out of the sandy bottom beneath it. The serious damage to the main house was of little concern to Cap at this time. It was habitable, and that's all that was necessary.

It was the damage to his fishing business that most worried him. What had to be cared for first? He needed a secure dock, although it could be a much smaller one, to tie the *Olive* to. Were the trap skiffs damaged? What of the entire supply of trap nets stored in the Ice House? What of the two traps that he had been unable to retrieve before the storm hit? Cap had to examine that situation first, as the barrels and netting might be hazardous to navigation.

He was able to contact three fishermen who lived nearby and who were willing to put their activities of recovering and rebuilding their own properties on hold in order to help my grandfather. With two work skiffs that had survived the storm without any serious damage in tow, Cap and his crew cautiously maneuvered the *Olive* around the floating debris in the channel and toward the seaward opening of the breakwater. The rollers were still heavy and high, but their surface was as smooth as glass. The water outside the Harbor of Refuge was very murky and probably would not clear up for several days or longer. That would be of little consequence, though, as there were no fishing boats in any condition to head out to the grounds.

The closest trap had been set on the offshore bend of the wall making up the major part of the Harbor of Refuge. So as Cap passed through the West Gap he headed due west for about five minutes, which would keep the *Olive* clear of any lobster gear hanging off and around the seawall. He then turned the boat in a southerly direction and slowed her headway, allowing time to turn one direction or the other should he encounter any storm debris or remnants of his fish trap.

And the latter is what he found when he came to the location where his fish trap should have been. Because the gear was on the southern side of the breakwater, it had taken the full brunt of the tremendous seas, tide, and wind, resulting in terrible devastation. The huge waves and the extraordinary flood tide had pushed the trap, barrels, twine, and even a couple of the huge trap anchors up onto the wall itself.

All that Cap and his crew could do was to slowly bring the work skiffs up as closely as they could and cut free what rope and twine might wash back out to sea, endangering any vessel venturing to close to the wall. One trap barrel still seemed to be secured to an anchor, so Cap tied off the *Olive* there while he and his men began the much larger job of hauling in what was left of the trap itself. The trap anchors, located by dragging a heavy chain back and forth across the bottom, were marked by dropping a flag buoy overboard on each one. They could be hauled up at another time.

The next day Cap steamed out to the last trap east of Point Jude. Surprisingly, it had survived the storm pretty well. Unlike the trap off the middle wall, this one had been in the lee of the Point Judith peninsula, which broke up—to a degree—the wave action that had been brought about by the hurricane.

Because the "parlor" had been laden with fish for the past week or so, most of the twine was in tatters, as the catch likely acted like a battering ram that was smashed against the twine by the waves. Otherwise, however, almost everything was intact. Cap figured that once the trap was hauled and brought into shore, it could be repaired with very little effort. For this he was grateful, for he was inundated with enough work to keep him busy into next spring.

It took many years for the local fishing industry to recover from the devastating hurricane of 1938. Lives, boats, and property had been lost, but the tenacity of the fishermen and their families helped them to survive, rebuild their fleet, and prepare for the new season that would soon be upon them. They went at the chore and succeeded.

THE FOG

"Time to tar the twine!" God, how I hated to hear Cap say that. It was almost always on the hottest day in July, and we would have to get a bonfire going under the huge tar kettle, which I called the "German helmet" because that's just what it looked like. If you could get hold of a World War II German soldier's helmet and turn it over so the opening was at the top, that's what that tar kettle resembled. Except it was as big as a small swimming pool.

When I was a boy at Snug Harbor, you see, there was no such thing as nylon twine; that didn't become available until years after World War II was over. The only twine we had for our nets, or rather, Cap's nets, was cotton. Untreated (untarred) twine would rot during hot summer weather in a day or two if wet nets were not hauled up on racks or otherwise spread out to allow them to dry. Even tarred twine needed periodic attention. Sometimes all we had time to do was drag the seaweed- and moss-covered nets up onto a sandy beach, and there they would dry out in the sun.

My job, before the tarring began, was to scavenge as much driftwood as I could, and that was no small task. It took tons—or, at least, what seemed to me to be tons—to even begin to heat up the "helmet." As it came up to temperature, "seven times hotter than it ought to be," Johnny Long, Jack Lewis, and I would open up the fifty-five-gallon drums of tar that had been brought to the site. While the intense heat was radiating upward from the fire, we had to pour the tar into the kettle—drum after drum after drum. It seemed that we would never fill that kettle up with enough tar to satisfy my grandfather. Between the fire under the kettle and the July sun beating down on my head, I felt like one of those Hebrew boys in Nebuchadnezzar's fiery furnace on the plain of Dura.

Once the tar was at a low, rolling boil, our next task would be to open up bales of new cotton twine and drag them over to a large wooden roller. Just before we began to feed the twine into the hot tar, Jack Lewis would tie a line to the leading end of the twine and pass it around to Johnny Long. They would then allow the rope and the twine to settle into the tar, swirl it around, and begin to haul the treated twine up out of the tar kettle slowly, as new twine was being eased into the tar. Thankfully, my job was to feed the untreated twine onto the roller as Jack and John pulled it through the tar and out of the kettle onto a contraption called the drying rack. This held the dripping twine up in the air, allowing the excess to run off, flow down a chute, and go back into the kettle. The treated twine was then stretched across another, longer rack that was closer to the ground; there it was allowed to dry for a few days before it was used to build new fish traps.

On this particular day, just before we took a break for lunch, the wind swung around from the southeast to the southwest, and the huge fogbank that we had been keeping our eye on all forenoon began to edge inshore. In just a half-hour or so we could not see Galilee, a few miles to the south of us. And by the time our break was over, the fog had settled in so heavily that moisture began to drip off everything around us. Of course this delighted me, as I was burning up with the heat of the sun, let alone the fire beneath the kettle. As the day drew on and we neared the end of our project, Johnny Long began to hum a little ditty that I'd heard a hundred times but could never understand, because he sang it with such a "New-fun-land" accent that only the melody was discernible.

I asked him what the ditty was about, and he said it was a chant to the fishermen lost in the fog off Newfoundland. Many a man had been lost as they were tub-trawling from their dories, their attention only on hauling in codfish, rebaiting hooks, and resetting their lines. These dorymen were supposed to keep a "weather eye" out for the anchored mother ship and where she lay in relation to their location. When heavy fogbanks moved in, they were to haul their trawls and row back to the schooner. But sometimes inattention would allow the fog to move in before they were aware of it, and at times neither the dory nor the man aboard were found when it eventually lifted.

The fog enveloping us was pretty heavy in my estimation, and I said so as I looked around. I could only locate the tar kettle by the red glow of the roaring fire beneath it. John looked up and said, "Dow! This ain't no fog at all. Let's finish up the work and I'll tell you how heavy the fog gets up in New-fun-land."

Well, you know, that stirred some extra energy in me and I redoubled my efforts to get the work done.

After we had finished spreading out the newly tarred twine, the day was pretty well over, so we went into Jack Lewis's shanty, which was close by. Jack made a pot of coffee so strong that you could have used it to tar twine. I scrunched down on the floor near Johnny's feet and waited for him to begin his tale, "The Fog."

It seems that John had signed onto a swordfisherman, the *Elvira*, a schooner of about ninety feet long. He said that the first day on the grounds had yielded fifteen swordfish, and the captain and crew were happy over the prospects for the days to come. The second morning brought a bright, clear sunrise and a calm wind, the best kind of day to see fish finning on the surface. Everyone was anxious to get started, so when the captain told the men to climb aloft, each one raced up the ratlines to get to his assigned place and, he hoped, to see the first fish of the day. John was located just below the man at the very top of the topm'st, and he began scanning as soon as he got settled into his hoop and secured for the day.

The skipper had been listening to the morning weather report from Newfoundland, and although heavy fog was expected for late in the afternoon, they would have the major part of the day to fish. And so they did, with no little success; the *Elvira* didn't do as well as on the first day, but she did well enough.

As expected, a massive fogbank appeared to the southwest, but the captain hoped to fish for another hour or two before calling it a day.

Everyone was caught up in the excitement of fish being harpooned one after another, and the captain, out on the stand as the striker, neglected to keep track of the progress of the fog. The *Elvira* was hours away from land, there were no other boats in sight, and she was well out of the shipping lanes, so he probably gave little thought to what appeared at a distance to be a typical "summer fog."

The skipper had just sunk the iron into a fish that he figured went five or six hundred pounds and, as the helmsman swung the stern away from the warp that was running rapidly overboard, the vessel turned into the southwest and right into the heaviest fog that any of those old salts had ever seen in all their days at sea. The captain called to the men aloft, telling them to be careful coming down the ratlines. Water from the fog was literally saturating their clothing, and they found it difficult to move their legs during their descent. As they grabbed the side stays, their hands became very slippery. It was as though they were actually overboard. One feller, a heavy smoker, yelled out, "I can't breathe. There is no air! I think I'm overboard! Somebody help me!"

John had not begun to get out of his hoop and was just unfastening the safety belt when he got slammed back against the mast by something much bigger than himself.

When he got his bearings, he was looking eyeball to eyeball into the face of the doryman who had picked up the keg of the last fish ironed, just before the fog hit. Both men were terrified, because both knew where each should be and neither should be able to look into the face of the other. Johnny, quick of wit, grabbed his fellow crewman by the arm and told him to secure the dory by tying the painter to the mast and instructed him to climb out onto the masthead crossbars so they could climb down the rigging together, as both were disoriented by their encounter. This they did at as quick a pace as they could muster, slipping and sliding as they took one step at a time.

They arrived on deck to find the crew huddled around Nick, the smoker, who had cried out for help. He had collapsed as soon as he reached the deck. The cook, who served as the ship's doctor, had rolled him over and listened to see if he was breathing. He declared that Nick was dead.

"He's not only dead, but he drowned—drowned in the fog," said the cook.

The captain took a head count to see if everyone else was safe. All responded as their names were called, with the exception of Nick, of course—he was just lying there, dead—and Leif, the ball man, who was nowhere to be found. John then told the captain about being hit by the dory while seventy feet off the schooner's deck and George, the doryman, confirmed his story.

The captain announced that the fog was too thick and too dangerous for anyone to begin looking for the ball man. "We have to wait until the fog lifts," he said.

Well, that took two days, and then the sun shone brightly again, burning off the remaining haze. The cook was on deck preparing Nick to be buried at sea.

He had just sewn the body into an old piece of sailcloth and had put in some lead weights from the tub trawls that had been put aside for the season when he heard a noise from above. Looking up, he saw the dory swinging back and forth, striking the mast as the *Elvira* rolled in the swell. The skipper sent John and George up the mast to lower it down to the deck "before it gets stove up." When they had climbed to John's hoop, they looked up above the dory and then called down to the captain on deck.

"We found Leif."

"Where is he? Is he all right?"

"Not really," John hollered. "He's been pinned to the mast by the fish you ironed as the fog hit us. The sword went right through his heart and it's stuck in the mast, sunk in about five inches, I'd guess. The fish is dead too."

"Well, ease them both down on deck," said the captain.

This was done in good order—first the swordfish, then the dory, and lastly Leif.

"Now," John said to me as I hung on every word of his tale, "*That's* the kind of fog we have up in New-fun-land. You see, the swordfish couldn't tell the difference between being in the ocean and swimming in the fog, the dory floated like it was on the briny deep, and poor old Nick drowned seventy feet above the deck of the *Elvira*. Have you seen anything like that around Point Judith?" he asked.

I slowly shook my head from side to side. "No, I never have."

Now, John's tale may have been fantasy, but the *Elvira* was a real swordfishing schooner, and the way she fished makes almost as good a story as "The Fog."

The *Elvira*'s forward mast—her mizzen—and its topm'st (topmast), an extension of that spar, stretched heavenward about seventy feet altogether.

When the crew was looking for swordfish, there would be seven men aloft. Six were in small, round hoops that kept them from falling when the ship rolled and thrashed about, especially while going to windward. The seventh was called the "ball man," as he rode the very top of the topmast. The only thing that held him aloft was a canvas strap about the width of a fire hose. He would settle his butt into that strap and then hang onto the top of the mast with both arms wrapped tightly around it to keep him from falling to the deck below.

The swordfish stand was a narrow platform that extended forward some twenty feet from the bow of the boat. The eighth man aboard—the striker, who used the harpoon—had to walk with death every time he went out to his position at the end of the stand and when he returned inboard. About fifteen feet of the ship's bowsprit was supported by two heavy wire stays. One ran from just below the crow's nest (about six feet from the top of the mizzen mast) down to the top of the ship's stem. A second went from the very top of the mast out to the end of the bowsprit. Finally, a third stay ran from the top of the mast to a point about four feet from the outboard end of the stand. A lifeline ran horizontally from the first stay out to the second, and then to the third. The striker had about eight inches of wood on which to place his feet as he held tightly to the lifeline overhead on his way out onto the stand, bobbing and weaving, rolling and heaving as the schooner rose and settled into the seas around it.

The striker was responsible for setting up all of the required equipment for that most important job of hitting the fish's back near the dorsal fin. This was a target of about one square foot, and it might be as much as ten feet underwater and crossing the stand at about five knots in one direction while the mother ship was going up to eight knots in another direction. To minimize glare on the water's surface, the crew had to be certain that the sun was behind the striker while they were coming onto the fish, and they would try to take the seas anywhere but head-on.

Before the crew started finding fish, the striker had to line up the wooden swordfish kegs, which were about eighteen inches long and thirteen inches in diameter. Each was strapped with line in such a way so as to prevent the kegs from coming loose of the long warp (rope) that connected the keg to the dart, or "lily," that penetrated a swordfish that was harpooned.

The swordfish warp was carefully coiled in baskets or tubs in such a way that when a fish was hit and started to run or dive, the rope would flow smoothly out of its container until all that remained was the keg. This, in turn, would be set up so as to fly clear of the boat and all its rigging on its way overboard.

In addition to preparing the kegs and warps, the striker had to run the lily end of the warps out to a point near the end of the stand, securing them, perhaps a half-dozen at a time, to one of the outboard stays with a well-tied slipknot. This would allow him to strike a fish, then quickly reach for another lily and place it on tapered end of the soft-forged iron rod that was fastened with a collar to his harpoon, a wooden pole of fourteen to sixteen feet long. Then the striker would quickly stretch the warp along the pole without even looking at what he was doing, and tuck a small loop of the line under a six-inch leather strap fastened lengthwise on the shaft. This would hold the warp tight to the pole, thus preventing the lily from dropping off the iron shaft when he raised the pole to strike another fish.

He had to do all this while keeping his eyes sweeping from port to starboard, because at times several swordfish would fin all at once, and one might rise to the surface right in front of or beneath the stand. The harpooner had to be ready to strike without warning.

Sometimes, when the schooner was steaming to windward, the striker would be ducked beneath green water every now and again, and he would remain wet for the balance of the day. Along in late August or early September, the North Atlantic can be a mite cold. I know; I've been there.

All the time that these activities were taking place down on the deck, another crucial bit of work was going on seventy feet above. The man on the ball and those in the hoops had to constantly sweep the ocean near the boat, looking for fish. Oftentimes a swordfish would "horn out," or show both its dorsal fin and the upper lobe of its tail. But many times a fish would come up near the surface to enjoy the warmth of the sun, yet its fin and tail wouldn't break the surface. I'll tell you one thing; it was a big mistake to let the captain see you, the ball man, with your eyes focused anywhere but underwater. It was a great temptation to look up when one of the other men on the topm'st yelled out, "There he is!" But just as the boat was being swung around toward the finning fish, you might run over several, much-bigger fish underwater.

The rest of the men, usually six in number, scanned the surface of the ocean, near and far, for the distinct fins of a swordfish. Actually, as mentioned above, what looked like two fins was really the dorsal fin, which has a unique curve that distinguishes the swordfish from any other fish in the ocean. But what solidifies that identification is its tail, which stands erect and moves from side to side with a stiffness not found in other finning fish. A shark might momentarily give you a start when his dorsal fin is first seen, but its tail will flop from side to side. Once a swordfish is seen, there is no question as to its identity, and I've never seen a good mast-header make a mistake. Even if the broadbill is completely underwater, the almost electric bluish-purple color of these fish is simply unmistakable.

In addition to the men aloft, there were others with important and often more dangerous work—the dorymen. Once a swordfish was harpooned, the crew member assigned to it was cast loose of the mother ship, whereupon he rowed to the keg and pulled it into the dory. The sooner that he did this, the more likely he was to keep the struck swordfish from pulling the keg out of his grasp or diving deep enough to implode it from the force of the water pressure.

The greatest danger faced by the men in the dories was when they began hauling in the warp and it suddenly went slack. That meant the swordfish had doubled back and was following the warp back to its source—often at high speed.

When that happened, the doryman had to quickly grab a handful of the warp that he had been hauling in and coiling into a basket between his legs, throwing the coils as far away from the dory as he could, because the next thing that was likely to happen was that the swordfish would come "a-hellin'" right up through that coil and would sail into the air, thrashing around wildly. Either that, or the fish would attack the dory from below. My brother was hauling in a swordfish near another dory one time when this happened. The fisherman did not throw the coils of warp out away from himself soon enough. The swordfish came right through the bottom of the dory and its sword drove up the inside of the man's leg, penetrating his gut and killing him instantly.

Another thing a doryman had to be careful about was keeping his feet out of the warp he was coiling, and making sure that his hands stayed on the outside of the rope so he could let go of it instantly. Failure on either count could cost him his life, as he could quickly be pulled overboard when the fish dove toward the bottom. Over the years, I heard about more than one such accident.

I learned a lot about the sea from listening to the stories of experienced fishermen like Jack Lewis and Johnny Long, even when they were telling tall tales such as the one about the thick fog off New-fun-land. Often I heard things that later helped me when I found myself facing elements of nature that could have brought harm to the boat or those aboard her.

Yes, sir. If you have chosen a life at sea, you want to be learning and remembering all the time. It could well mean the difference between living and dying.

WALL OF FIRE

I remember movies being my family's favorite choice of entertainment. At the end of the day, my grandfather was usually weary from running his large and very profitable fishing operation, and movies were a way for him to get his mind off business for a while. But unlike me, he was able to walk away from the plot with nary a second thought.

Most films made during that period were centered on the war. The heroes were clearly defined and, of course, so were the villains. The action pitted good against evil. This simple formula could be projected on the screen in an exciting way that made an impression, especially on the mind of a young boy.

One such movie featured a daring military plan by American forces to destroy a Nazi U-boat base that had been built into a huge cliff abutting the English Channel. The enemy had enlarged a natural cave at water level but had left the opening so small that the U-boats had to enter at periscope depth.

Once through the entrance, the subs could surface in a tremendous cavern that could harbor all of the U-boats in that sector of war. Also within the facility was all the necessary repair equipment, including a marine railway and locks big enough to haul the subs out of the water for needed repairs, as well as torpedoes, food, and fuel for outbound boats. Needless to say, the fleet maintained at this enemy base was raising havoc with the U.S. merchant ships carrying supplies to England.

Every Allied effort to destroy the U-boat refuge failed. The natural protection of overhanging cliffs prevented bombers from hitting the target, and warships were unable to penetrate the small opening with their shipboard gunfire. But strike up the bands, for here came the marines: John Wayne, George Brent, John Hodiak—it didn't matter. These guys would succeed. Everyone in the theater knew that; what we didn't know was how they would do it.

This is where I began to sit on the edge of my seat, as though those few extra inches would put me right into the war room of the Allied forces as plans were developed. The marines, under the command of Hollywood's most handsome and undoubtedly bravest hero, came up with a brilliant plan: The Allies were to send in warships at night to offload a flotilla of fifty-five-gallon drums filled with highly volatile aviation gas. Once overboard, the drums were to be lashed together and, as the tide started to flood, they would drift toward those ominous cliffs. Working in the water under the direction of their dashing and very brave commanding officer, the marines would maneuver the long string of fuel drums toward that ever-so-small opening. Everything had to be timed just right, because if it took too long for the drums to reach the cave opening, the flood tide would raise the water level in the opening, preventing them from entering the U-boat refuge.

Another factor was that the last drums to enter the base were to have explosives strapped to them. When all drums were inside the compound, the explosive charges were to be detonated. This would cause a holocaust of fire that would set off all the rest of the fuel drums; then the resulting wall of fire would engulf the German submarines, their crews, and the replacement torpedoes. This, in turn, would destroy the enemy base for all time.

Of course, the explosives did not detonate when they were supposed to, and the second most important hero, the star's best friend, swam inside the U-boat base to ignite the drums, sacrificing his life for the victory of the Allied forces.

The curtains closed. We slowly left the.theater, everyone talking about the exciting scene they had just witnessed. Soon all were on their way home to sleep the night away. But not this boy.

My mind was racing. I relived each and every reel of action shown on the screen. Could such a feat truly be accomplished? Could a wall of flame really do the damage shown in the movie? Would it look as it did in the film? Would I be able to do what the commander's best friend did—sacrifice myself for the good of all mankind?

Such questions filled my thoughts for hours, until pure exhaustion finally brought sleep. It was short-lived, though, as my job each morning was to rise at 5:00 A.M. I was assigned to take heavy five-gallon cans to an old-fashioned gas pump that had to be hand-cranked. There was a flat, notched rod that marked off each gallon as you pumped. The crank was so far off the ground that I had to stand on a heavy box to get high enough to gain sufficient leverage to wind the handle over and over as gas flowed out of the hose into each empty can.

When all the gas cans were full, I had to lug them down the dock to my grandfather's trap boat, the *Olive*. On the forward deck, just ahead of the wheelhouse, was a thirty-gallon galvanized tank into which the gas I had just pumped had to be poured. This was no small task, as I was not much taller than the cans themselves.

On this special day I filled an extra can of gasoline and carried it up forward of the wheelhouse—but not to dump it into the fuel tank. On this day I was going to see if that story in the movie was real. Would gasoline really burn on water? Would it explode into a wall of fire? Is it possible that it could drift into other ships, destroying everything in its path?

The entire crew of the *Olive* was ashore in my grandfather's kitchen, enjoying a breakfast cooked by Grammy. They would dally as long as Cap would let them, downing as much grub as they could eat, for my grandmother was an outstanding cook.

I took advantage of this quiet moment, and as the morning sky was beginning to brighten into a beautiful day, I looked around one more time. There was no one in sight. I dumped the full five gallons of gasoline overboard. As soon as the last drop hit the water, I lit a wooden match and dropped it into the pool of gas.

Whoosh! Sure enough, the gas exploded into the fearsome wall of flame shown in the movie. It was huge, and only one thing saved the *Olive* from becoming engulfed in the fire as the Nazi subs had been: The tide was ebbing, and it quickly carried the holocaust away from the boat, and the flames slowly drifted down the channel. Oh my! The wall of fire was going to drift into the fleet of fishing boats down in Galilee and Jerusalem. I was going to destroy the whole fleet of my friends and family! Oh God! What could I do?

Before I could call for help, the flames began to dissipate. About halfway down the channel, the fire burned itself out. The fleet was safe, the *Olive* was safe, and no one else had seen the wall of fire, so I, too, was safe. Never in my young life had I felt such relief.

I brought all of the empty gas cans back up the dock to be filled again the next day, then I went into breakfast. As quietly as possible I sat down in the midst of the men, who were talking and joking with each other. I could not believe it. None of the crew had witnessed my foolishness. There was no teasing to undergo; there were no embarrassing questions to answer.

I could not believe that I had gotten away with such a stupid stunt without anyone's being aware of it. Slowly I calmed down, and just as I was about to enjoy my bountiful breakfast, my grandfather leaned over and whispered into my ear, "I'll have to dock your pay for the gasoline."

Yes, young people are affected by what they see, hear, and read, so be very, very careful about their entertainment.

ENCOUNTER WITH WHITE SHARKS

11:15 P.M. July 1, 1940, Jerusalem Beach, Flood Tide

The sky was coal black, marked only by ribbons of a waning moon. The wind was out of the sou'west, lying gently across the three men as they stood waist-deep in the fast-flowing easterly tide. All three worked together in a mill in Providence.

They had been waiting all winter for this night. They had been buddies since high school and had fished together in every body of freshwater found in Rhode Island and some miles beyond.

Every coffee break they took, every movie they had seen together, just about every time they were together brought on the conversation about their plans to go to Jerusalem to catch giant striped bass. Now here they were—the self-proclaimed Scituate Striper Club.

The beach they were fishing was known for a fast tide, and a number of sluiceways, or gullies, had been carved into the sand by that tidal action. Someone fishing alone and moving along the beach while following a school of fish might well fall into one of the deep spots and, having no one to help him, might well drown. These men were aware of this danger and had taken steps to provide as much protection for themselves as possible. They tied themselves together with a light nylon rope, and because Scott Taylor was the biggest of the trio, he was assigned to be the anchor for the other two.

This would also allow the two lighter men to venture a few feet farther into deeper water where the bass were most likely to strike. When all three men were in place, they bent on the "blue mullet" plug designed and tested by Herb Irish up in Wakefield. This lure had caught more stripers at this time of the year than any other on the market. Having planned well for this trip, they each had several in their kits.

Within minutes all three had hits. Howard Maney was the first to land a fish. "Man! It must weigh fifty pounds. What a fight! Hey, guys, this is what we've been waiting for!"

Scott was next to haul in one of those big ones, yelling with excitement as he did so. In his anxiety to catch his share, Wayne Barrows, the smallest in stature of the three, took a couple of steps out into deeper water and soon felt the strike that he had been waiting for.

"I've got 'im!" he yelled over his shoulder. "He's running offshore on me! Hold onto me, Scott! It's all I can do to hang on!"

Just as he called out, something hit the backside of Wayne's legs and threw him into the air, churning the water all around him. He disappeared under the surf. Scott called out to Howard, "Help me! I'm losing my hold on Wayne! Quick, grab hold of his line with me!"

As soon as the two men threw their weight back against the lifeline and began to dig their feet into the hard-packed sand, Wayne broke the surface. Although he was sputtering and vomiting a gut full of salt water, he was still hanging onto his rod and reel. The other two men quickly dragged him up onto the beach and then ran down to him.

"You all right? What happened, anyway?" they asked, almost in unison.

Wayne was exhausted, but replied, "I've still got the fish on the line. I feel some weight but there's no fight. Hold on while I reel him in." All three watched anxiously as the line came in. When the plug broke the surface, hooked to it were the head, gills, and a third of the body of a striper twice the size of those caught by Scott and Howard.

"What in the world could have taken so much of that fish in one bite?" asked Howard. "It must have weighed seventy-five or eighty pounds."

Wayne spoke with a tremble in his throat: "That's what knocked me down, and just as I went under I could see a large, dark bulk. It had to be a shark."

"No shark that I've ever seen or even heard of around here is big enough to do what this one just did," said Scott. "Fellas, we've had enough excitement for our first day," he continued. "What do you say we go back to the cabin and turn in? I've arranged for us to go out with Cap Clark to watch him haul his fish traps in the morning."

"What time in the morning?" asked Wayne wearily as he hauled himself to his feet.

"Cap said that he's making two trips tomorrow," answered Scott. "He's hauling a couple of traps east of Point Judith first, and he told me that he should be back in around nine. If we're down on the dock by nine-thirty, we should be fine."

As the other two moaned and groaned, Howard said, "I don't know if we can handle this kind of excitement for the next two weeks." At that, the men gathered up their gear, iced down their fish in the back of the pickup, and drove to Snug Harbor.

7:30 A.M. July 2, 1940, Jerusalem Beach, Ebb Tide

The sliding-glass door opened in the cottage overlooking the ocean and beach as six young girls ran into the sunlight, yelling and laughing as children do at play. They were all carrying some kind of beach equipment—blankets, towels, and folding chairs—and they were anxious to stake out their territory before the folks from the city came roaring in. Experience told them that this holiday week would draw thousands to this very beach, as well as others in the area. Only those first on the sand would get the best selection of spots and, strangely enough, their obvious boundaries would be recognized and honored by the waves of beachgoers to follow.

As soon as their gear was strategically placed and their ownership declared, the girls ran bouncing and screaming into the southerly ground swell that had made up the night before.

They were diving, swimming, and playing in the exact spot where Wayne had encountered "something" just a few hours earlier. The water was cooler now due to the ebb tide, whose current ran from east to west, eventually entering Long Island Sound. This flow often brought in mackerel, butterfish, and skipjacks—bait*fish.

Suddenly, one of the older girls cried out, "Look! A seal or something is rolling around and thrashing the water. And there's some little fish jumping out of the water. Let's go see what it is."

In unison the others yelled, "Okay, let's race to see who can get there first."

One of the smaller girls, who happened to be the best swimmer of the three, reached the area first and dove to see the seal. Something big and dark was just edging out into deeper water.

"We've missed him," she reported. "The seal is gone. I've never seen one in the ocean before. I wish I could have seen it feeding."

With the excitement over, their attention turned to a beach-ball game one of the slower swimmers had begun. "Let's see who can throw the ball the farthest," she called to the others. With that she tossed it out to sea, and the ball hit the water with a splash. Immediately, something snatched it and pulled it beneath the surface.

The oldest girl cried out, "What happened to the ball?"

"I don't know," answered the first girl. "Everything happened so fast I couldn't tell what it was."

Just then a bell began to ring, and a woman at the cottage called the girls in for breakfast.

10:45 A.M. Monday, July 2, 1940, West Breakwater Wall, Inshore Trap, Ebb Tide

The *Olive* was towing a string of work skiffs and had a dozen crewmen aboard, along with a half-dozen "city folk" who had requested the opportunity to go out and watch fish traps being hauled.

A few of the visitors were laid out on the forward deck, rolling and retching and moaning from seasickness. Wayne, Howard, and Scott were hanging onto the side stays of the mast, trying their best to stay upright as the *Olive* lifted and rolled and settled into the swell that was rising from the sou'west. They were just one belch from upchucking themselves, and they were working as hard as they could to keep from doing so.

Just as they were about to lose the battle, Cap called out, "I'm going to skip the offshore trap; we'll haul the smaller one first." This trap was some three hundred yards off the beach and nearly in line with where the three young girls had been playing.

Without any specific directions, the crew reacted as quickly and smoothly as a machine. As soon as Cap threw the engine out of gear, they hauled the skiffs alongside the *Olive*, and two and three at a time jumped into them and pushed themselves away from the mother ship, heading toward the far side of the trap.

Cap then made a wide turn offshore and circled back to the trap. He wanted to have the frame on his port side. He slowly brought the *Olive* alongside the gear, and I grabbed the lanyard that was tied to the nearest trap barrel. I then brought it up to the stern bitt and secured it. I worked the trap frame slowly out of the water and secured the midsection to the *Olive*'s quarter bit with another lanyard.

My next job was to untie Cap's work skiff and haul it up alongside. Normally, I would stand on deck with my knees against the bulwark and lean into the weight of the skiff, but today I had an audience. I jumped up onto the caprail of the bulwark, which was about six inches wide and had two pieces of steel half-round running along both edges for the entire length of the cockpit area, about twenty feet.

I forgot that we had hauled three other traps earlier in the morning and that the caprail was covered with seaweed, fish guts, and seawater. My balancing act lasted about twenty seconds, and then I plunged overboard—into the trap yet to be hauled.

In the few seconds that I was airborne, two very distinct thoughts bounced around in my youthful head: "You can't swim," and "Grab something before you go under the mackerel, squid, and whatever else might be in there." According to my mother, I've been to sea since I was six weeks old and have never had a fear of water, so I have no excuse for not being able to swim. But it was definitely too late to think about that.

My entry into the cold Atlantic that morning was definitely not graceful. I hit the surface on the flat of my back and quickly sank toward the bottom of the net. I reached out, stuck my fingers into the mesh of the twine, and held on, slowing my downward trajectory. Then I began inching my way upward.

Just as I broke the surface of the water, gasping for air, something rammed into my legs, driving me upward. This unexpected assist allowed me to grab the caprail of the *Olive* with one hand while my grandfather took hold of the scruff of my neck. Reaching over, he slipped his other hand around my belt, then in one heave he hauled me aboard the boat and dropped me unceremoniously onto the deck. Without a fare-thee-well he turned toward his skiff, which was now thumping against the *Olive*, and as he lifted one leg over the bulwark he turned to me and said, "Come on boy, we're wasting time."

I stood up, emptied the water out of my boots, put them back on, and staggered to join him in the skiff. Nothing gave me more pride than being able to work beside my grandfather as we hauled trap. His apparent coolness to the dunking incident didn't surprise me. Cap had been sailing from the age of eight, when he'd served as a cabin boy on square-riggers. In his years before the mast, he had witnessed men falling from the yardarms—sometimes overboard and sometimes onto the deck. He had been shipwrecked on an uncharted island and survived storms more powerful than anything seen in our waters. He had experienced the loss of his vessel during a storm in the Pacific, when he spent twenty-eight days in a longboat before being rescued by a homeward-bound whaling ship. Yes, J. E. Clark was not one to panic when a young boy simply fell overboard off Matunuck Beach.

Gramp and I sculled across the width of the trap toward the leader that extended to the west wall of the breakwater. As a variety of fish worked along the beach or came in from deep water to chase the baitfish in and around the riprap, they encountered the mesh in the leader. They would follow the leader offshore, looking for a way to get around it, until they hit the wings of the trap.

This would confuse them, sending them swimming in circles, looking for a way out. The schools would eventually find the opening that led offshore from the wings; flowing through it, they would enter the section of the trap called the kitchen. Here they would again begin to circle until they found the second opening, again in the offshore side of the trap. Sensing open water ahead, they would rush into the final and most captivating section of the trap, the parlor.

Nearing the offshore end of the leader, Cap turned his skiff, allowing me to reach overboard and pick up the twine of one of the wings. As we hauled it over the rail, a couple of men in another skiff were doing the same thing on the opposite side of the leader. When this was accomplished, the opening to the main body of the trap was closed, preventing the body of fish from exiting the encompassing twine.

Then, spreading out along the barrels supporting the perimeter of the kitchen, the men in all the other skiffs began to haul the twine of that enclosure, allowing it to go overboard behind them as they worked their way toward the parlor.

All the fish still in the kitchen would scurry toward the last opening ahead of the advancing net. Once the line of men crossed the kitchen, they would reach down to untie the flaps of twine that would close the aperture in the last set of wings.

Now the only way for the body of fish to get out would be over the floats holding the twine in place. I've seen that happen, but that's another story.

Now, positioned in nearly a straight line, our skiffs bow-to-stern, we began in unison the most difficult task of all: "hardening up" the mass of trapped fish while passing the twine through our hands. At the same time, we had to hold as much strain as possible on the net to prevent a surge of fish from hitting a loose section of twine, causing a tear and allowing our catch to escape. As the fish were concentrated in a smaller and smaller area, and as we came closer to the *Olive*, the biggest of the men in our crew would grab the large scoop nets and begin to bail whatever valuable species we'd caught onto the deck of the mother ship. Some of the crew would jump from their skiffs to the *Olive* and use picks to throw overboard the unmarketable trash fish.

This hardening up, bailing, and picking would go on until the last fish in the trap were scooped aboard the *Olive*. Then the men remaining aboard the work skiffs would reverse the hauling procedure, dropping the twine back overboard until all sections of the trap were in the same position as before we started. The wings were again tied back, waiting for the next schools of fish and whatever else might grace our inviting fish trap.

On this particular day, though, emptying the trap required a lot more than scoop nets. Shortly after the bailing process began, the water commenced to roil and thrash, spraying all hands, as well as the sick and quasi-well city folks. Everyone began to shout in excitement, but no one could see just what the source of the commotion was. There would be a flash of blackish-gray skin, but no evidence of a head or tail.

Was it a porpoise, a seal, a small pilot whale? All of these had been entangled in our nets at one time or another, but they had been quickly identified, even before the hardening of the twine took place.

It took all the muscle of our very strong crewmen to bring our mystery into view. When I saw what had been in the net with me, and what had evidently rolled up under me and boosted me back aboard the *Olive*, I began to shiver with fright. There were two of the biggest sharks I had ever seen, their jaws snapping wildly. Cap was anxious to prevent them from destroying the trap so he went into the wheelhouse, grabbed a shotgun and some shells, and quickly ran over to the side of the boat. Making sure that all hands were out of harm's way, he fired two shots into the head of each shark as they rose to the surface. When they had ceased thrashing, Cap put a line around the tail of one shark and, with the help of a crewman, hoisted it to the deck of the *Olive*. Just as quickly he brought the second one aboard.

Once on deck, the sharks again began to thrash and twist with their giant jaws snapping. Cap put a knife to a stone, then went to each one and deftly opened its gut. As the contents spewed out we could see the partly digested body of a seal in the gut of one shark; the other yielded a large section of a striped bass and the tattered remains of a beach ball. Both were filled to overflowing with mackerel, the bulk of our catch this day, and both had an assortment of tin cans, bottles, and other flotsam jammed into their huge intestines.

Cap Clark, the author's grandfather, shows off a huge white shark caught in one of his fish traps.

Once the fish, sharks, and remaining crewmen were aboard and the work skiffs were tied aft, ready to be towed, Cap put the *Olive* in gear and turned onto a course toward home. He never made mention of the near disaster of my being overboard with two of what we learned later were white sharks—man-eaters.

Shortly thereafter, the three men from Scituate told Cap about their adventure of the night before, and all of them agreed that they had encountered these very same sharks. As the story spread over the bulkhead and around Salt Pond, the parents of the three young girls came to the same conclusion about what their daughters had thought was a seal, especially since the tattered remains of the beach ball had been digested by one of the huge creatures.

Upon their arrival at Snug Harbor, and after pictures of the sharks were taken, Howard Maney, Scott Taylor, and the still-shaking Wayne Barrows could not wait to tell everyone about the "one that got away." They spent the balance of their vacation in various gin mills recounting and expanding on the story, and fully enjoying their fifteen minutes of drama.

As for the huge striped bass they wanted to catch—well, there was always next year.

SUDDEN DEATH

It was the last week in June 1942; the war was everywhere—except in Snug Harbor.

The day was going to be a hot one. The early-morning mist, which usually hung over the waterfront on a summer day, had already burned off. We had hauled the trap off the west wall, and it was full of scup. We had just relieved the *Olive* and two work skiffs of the first load, and we were getting ready to go out for a second one. My grandfather asked me to move one of the empty skiffs. She was wedged between the *Olive* and the dock, where two of the crew had been bailing the fish onto a chute. Here we culled (sorted) them, picking out the small scup and any sea bass or other marketable species that were mixed in.

The *Olive* was lying with her bow heading south, so I had to let her stern lines go and then work the skiff around the transom, re-secure the dock lines, and pull myself along the offshore side of the thirty-foot trap boat. At first, the task was relatively easy, as I simply stood on one of the seats in the skiff and leaned over to work hand over hand. But my job became more difficult as I approached the bow. I was starting to feel pressure from the ebb tide, and, because of the rise in the *Olive*'s sheer line, I had to reach higher up and move the skiff while hanging on with just my fingers.

Here I was, an eleven-year-old boy, stretching my arms overhead while trying to keep my little boat from scudding away in the tide, and there was no one to help, as the rest of the crew was bailing out fish.

I was reaching up higher and higher, my neck craned up and backward, when I saw two navy planes from Quonset Point crash head-on in one huge explosion. I screamed to the crew, and as they looked heavenward, the tail section of each plane fell from the mass of flames and smoke and twisted into a spiral toward Earth. This was followed by the huge, molten wreckage of the rest of the two intertwined planes. There were no parachutes. Just smoke and fire. As the planes fell, there were a number of smaller explosions, one after another. We learned later that these were from the cannon and machine gun shells being ignited by the heat.

At the time, however, I just kept hollering for help; in the excitement I had lost my handhold on the *Olive*. Jack Lewis jumped aboard the big boat, and I tossed the skiff's painter to him. He threw a couple of half hitches around the *Olive*'s bow bitt and then reached down to haul me aboard.

By that time everyone was scrambling for trucks and cars in which to head for the crash site, about a mile up the road from Snug Harbor. Jack and I jumped into Thad Holburton's pickup with him.

The two planes had crashed into a huge gravel pit recently excavated for the new Jerusalem road, which was to run down to the docks on the west side of the harbor at Galilee. The site of the crash was the most frightening scene of death and destruction that I had ever witnessed. Just as the grown-ups were about to run up to search for survivors, the crash site began to resonate with rapid fire from the weapons aboard the planes. One of the crewmen hollered, "Hit the deck!" and we all dug down into the earth in an effort to escape ricocheting bullets.

While we were hunkered down behind anything that provided protection, waiting for the shooting to stop, the local volunteer fire department showed up. Three or four firemen, hoses in hand, came around a high bank of gravel that blocked their truck from the horrific conflagration. As a few fifty-caliber shells careened off the sides of the gravel bank, the firemen, too, dove for cover behind some huge granite boulders, one squirming as deeply as possible into the rocky gravel. All of them were trying to lower their silhouettes in an attempt to minimize the chances that they'd be hit by a stray round. They did have the presence of mind though to yell back at the firemen near the trucks, "Turn those hoses on full force, NOW!"

The hoses were aimed high so that the stream of water arched over and began to hit the remains of the planes. It took upwards of a half-hour for this flow to begin to subdue the flames and smoke, because the hoses were sprayed in a general direction, aimed only by the hiss of molten aluminum as water hit its target behind a screen of smoke. Gradually the explosion of shells slowed and then ceased. No one was sure if the water was that effective, or if the guns were simply running out of ammunition. (It later proved to be the latter.)

As things began to quiet down, a few of the men slowly raised their heads, and when they felt certain that the danger from the guns was over, they called out, "You can get up now; it looks safe. Let's look for the pilots."

As any normal, obedient boy would do, I outran everyone to the crash site. I sure wish that I hadn't. I was the first one to find the bodies. There were two men in each shell of a plane. All were charred to the point that it was hard to identify the remains as human. Each man had his right hand burned into the same position: right arm across the chest and the right fist next to the left shoulder.

"What's wrong with these men? They're all burned into the same shape," I yelled at no one in particular. Thad worked his way up to my location and quickly said, "They were reaching for their parachute rip-cord, but they didn't make it. They died before the canopy opened."

It was then that the condition of the bodies and the horrible smell and the thought that four men had just died within my very sight, caused such revulsion that violent sickness overwhelmed me. I puked my guts out and then started crying and couldn't stop.

Jack Lewis came over, put his arm around me, and said, "Come on son, there's nothing more to be done here; let's go home."

I was a long time getting over my first experience with death in the field. I had seen dead people before. My father's mother, and, later, his father were buried from our living room (now there's an oxymoron), and although I was uncomfortable looking at them each time I had to come downstairs from my room, death was more sanitized, more distant, if you will. Those men in the plane wreckage were far from sanitized, and they were definitely up close.

It was the last week of June 1942; the war was everywhere—even in Snug Harbor.

THE SPIES

Snug Harbor, you'll remember, was about five miles up Salt Pond from the breakwater at Galilee. Due east was Point Judith lighthouse and the East Gap, another entrance to the anchorage behind the middle wall.

Southwest of Point Judith and the breakwater lay Block Island, and due west of that was Long Island, New York. Well, that layout of land and sea made a natural waterway where merchant ships could sail in relative safety, because they had the mainland of Connecticut to their port side and Long Island to their starboard all the way from New York City to Montauk Point. But there it got a bit dicey. They had to deal with open water from Montauk to Block Island, then again from Block Island all the way to the Cape Cod Canal. Several Italian and German submarines had been sighted in those areas of open water. In fact, a few had been sunk by bombers flying out of Quonset PointNaval Air Station, just a few air miles north of Point Judith.

When the war began, or shortly thereafter, the military bought up most of the land at Point Judith, which was a peninsula jutting into Narragansett Bay. Then they built a large fort with underground tunnels running from one gun turret to another. Each of the turrets held a single sixteen-inch cannon mounted on swivels, so they could cover the area from Block Island to the east, toward Martha's Vineyard and the entrance to the Cape Cod Canal.

There was only one problem, and it was a big one: When the fortification was complete and a day was designated for test-firing those big sixteen-inch guns, military officials sent a notice to the people in the surrounding homes as to when the test-firing would take place. At the appointed hour, they fired the guns, first singly, and then rapidly, and then all at once. Well, Mister Man, the concussion from those guns broke nearly every window, shook down most of the old chimneys in the farmhouses, and upset the cows, laying hens, and people for eight miles around. There was such an uproar from the folks in the area that the military never fired the guns again. At the end of the war, when they were dismantled, they were almost as good as new.

All this is to help you understand that the good people of Galilee and Jerusalem and Snug Harbor were no strangers to the activities of war.

The Coast Guard used to walk the beaches at night because submarines let spies off in rubber rafts, so what happened to me was not so unusual. I was bailing out my grandfather's work skiff, as well as all of the other skiffs, because we had recently finished unloading the fish we had caught in the traps set outside the breakwater walls. In fact, we had caught tons of squid, and all the skiffs were covered with ink. If that happens and you don't wash everything down quickly, the ink will dry and smell to high heaven. That was my job, and it was no small task. My grandfather's skiff was only about fifteen feet long and six feet wide, but it sculled like a cut cat. It was my favorite boat to use, and Gramp would let me have it whenever I wanted. The other fishermen's skiffs were much bigger; one that I called "The Battleship" was about twenty feet long, double ended, and bellied out amidships. It looked just like a battleship and was even painted battleship-gray all over. (Every boat and skiff that Gramp owned was painted gray down to the waterline, but then the bottom planking was covered with "red lead.")

Because I was busy finishing up the bailing—which, by the way, was done with a big grain shovel because it was the fastest way to get the job finished—I had not noticed the small ketch dropping anchor out in the channel just east of our dock.

I heard someone hollering and looked up to see three men waving to me as if they wanted me to come out and bring them ashore. That was a common thing in the summertime, what with all the yachts around; hardly any carried dinghies. So I waved to let them know I would be right along, but I yelled back that I had to finish bailing out the skiff because there was still about two inches of fresh, clean seawater in her from my wash down. Now this is the strange part: They yelled back that they were in a hurry, asked if I could come right out, and said that they would pay me a couple of dollars if I did so. I let the painter go and sculled right out to their sailboat.

All three men were dressed in slacks and white shirts with their sleeves rolled up a bit, and all were wearing brand-new dress shoes. That caught my eye. There was rationing all over the country, and shoes were nearly impossible to get. Everyone I knew wore scuffed and well-worn shoes, and few were even polished. Now, I told those fellers to wait just a few more minutes while I finished bailing out the skiff. No sir, they jumped down into the two inches or more of water, *salt water*, in those brand-new shoes.

Now, you just don't do that in front of a young boy who has seen every war movie that was ever made. I knew as sure as shootin' that these fellers were spies, German spies for certain. Then they gave me two brand-new one-dollar bills—another mistake.

Salt water does things to clothes and shoes and money; nothing stays new for more than a day or two. Those men had not been to sea for very long. They said that if I would be available to take them back aboard in about four hours, there would be another two dollars in it for me.

Well, as soon as those guys took off up the road—hiking it all the way, I guess; I didn't see any car pick them up—I scrammed right into my grandfather's and told him the story.

The way I made it sound, he got a little suspicious, as well, and called the Coast Guard down in Galilee.

About half an hour later, the picket boat came up alongside the ketch, and several men climbed aboard. Some were in Coast Guard uniforms and some wore slacks and shirts. I bet those men were from the FBI. I headed right out to the ketch; they told me to keep away, but I just drifted off a few feet from the sailboat.

I could still see what they were doing topside. Two of the guys in street clothes began to take the hinges off the companionway hatch, and that concerned me. Anyone knows that if you disturb paint around hinges, the damage will be spotted right away. Those spies were going to see that someone had been on board and had even gone belowdecks, and they would think that it was me. Boy, was I in a lot of trouble now.

I could hear some of the Coast Guardsmen throwing things around down in the cabin, but their voices were muffled and I couldn't tell what they were saying or finding. I didn't know how those men were going to fix things up so no one would know they had been aboard, but they wouldn't let me hang around any longer to find out, so I headed back to the dock. Those guys made me real nervous because I had no idea how they could make things right before the Germans came back. At least, I *thought* they were Germans.

Within the hour all of the Guardsmen and FBI agents climbed back into the picket boat and came over to our dock. First they spoke with my grandfather, and then they talked to me. They asked me if I was up to bringing those strangers back to their boat without getting scared. Were they kidding? Would John Wayne get scared? I reassured them that I could handle "this dangerous assignment" without a hitch.

A few hours later, the men in the new shoes showed up, and sure enough, I was available to ferry them out to their little ketch. Man, was I nervous as they climbed on board, but as far as I could see by looking over the rail, everything was shipshape and in "Bristol Fashion." No one seemed to be suspicious. I still don't know how those repairs were made, but I didn't want to hang around to find out, so I took the two dollars they owed me and swung away as fast as I could. They up-anchored and sailed down Salt Pond and on out through the West Gap in the breakwater. No one stopped them. What was wrong, anyway?

I had laid my life on the line, and the Coast Guard was letting them get away! I grumbled about things for a few days but was beginning to put it out of my mind when a Jeep drove up with a couple of naval officers in it. They wanted to speak to my grandfather and me about the incident. *Incident!* It was the turning of the war as far as I was concerned.

They told us that those men were indeed German spies and that they had been picked up by the Coast Guard offshore of Point Judith. They were apparently hoping to meet up with a U-boat later that night; then they would have scuttled the sailboat.

Just as in the movies, the good guys won, and all was right with the world—at least for a few days. It was just another exciting episode in the life of a boy in Snug Harbor, but it was a day I'll never forget.

U-853: A CROSSING OF PATHS

Author's Note: This story is based on the recorded maneuvers of the U-853*, reports of the rescue-boat crew, research by experienced seamen, and data from German archives. The dialogue of members of the submarine's crew is supposition, as there were no survivors.*

The end of the Great War was nearing in 1918 when the keel of the coal collier SS *Black Point* was laid in a shipyard in Camden, New Jersey. The ship was 369 feet in length, measured 55 feet on her beam, and she was designed to carry 7,500 tons of coal. Her owner was the Sprague Steamship Company of Boston, Massachusetts.

As the vessel slid down the ways during her launching, it never entered the minds of the firm's principals that she would have anything but a nondescript history, lugging coal from one port of call to another along the Atlantic seacoast. Yet this modest ship would go down in the annals of naval warfare, to be pondered by historians forever after.

Twenty-four years would pass as the *Black Point* sailed peacefully along on her prescribed duties. War clouds had started storming again over Europe. Germany began her efforts to build an invincible naval force to rule the Atlantic and beyond. On August 21, 1942, at Bremen, Germany, the keel was laid for a submarine to be christened with the uninteresting name of *U-853*. The boat was 252 feet long with a beam of 23 feet and a height to the top of her conning tower of 31 feet. She was a type IXC/40 Class U-boat, with a displacement of 1,545 tons. Her armament consisted of two twin 20mm anti-aircraft guns, one 37mm anti-aircraft gun, one 105mm deck gun, and six torpedo tubes. She was commissioned on June 25, 1943, under the command of Captain Helmut Sommer.

The *U-853* made three patrols under three captains, and in October 1944, while under the command of Captain Gunther Kuhnke, she took her first prize, sinking a merchant vessel of 5,000 tons. This earned her skipper the Knights Cross and a promotion to Commander of the 33rd Flotilla. It was after this victory that the sub got the nickname *Der Seiltaenzer*—Tightrope Walker.

The destinies of the *Black Point* and *U-853* were about to converge.

On October 1, 1944, *U-853* was assigned a new commander, Oberleutnant Helmut Froemsdorf. He had a crew of fifty-five seasoned veterans, many having sailed on this boat since her commissioning in 1943. Their area of patrol was the eastern seaboard from Canada to the New Jersey coast, and she arrived at her operating position late in the month of April 1945. Before leaving home port, the captain had been informed that the war was going badly for Germany, and Froemsdorf was not certain if he would see his family and country again. In fact, the bad news was confirmed when he met up with the rest of the wolf pack off the coast of Nova Scotia. Captain Froemsdorf discussed future action with other U-boat captains during his last night with his old seafaring friends, and many said that they would scuttle their boats before they would surrender to the Allies. Word had been passed along that several U-Boats had done just that north of Scotland, with all hands aboard.

On April 23, 1945, the subchaser USS *Eagle* (PE-56) was sunk by a German torpedo off Portland, Maine, killing fifty sailors. The official navy record claimed that a boiler explosion sank the vessel. Later, however, declassified information showed that a German sub (*U-853*) was in the area. Thirteen men survived the *Eagle* ordeal, and two are still living as I write this. With the record corrected, all those aboard are now entitled to the Purple Heart.

U-853 had made her second kill. She was now on her relentless prowl south along the coast, looking for another target.

On May 1, 1945, Hamburg radio announced that Adolf Hitler was dead. Grand Admiral Karl Doenitz took over as Der Fuehrer and immediately began to arrange a surrender.

On May 4, 1945, with World War II in Europe coming to an end, Admiral Doenitz gave the following order. "All U-boats, cease fire at once. Stop all hostile action against Allied shipping. Doenitz."

Froemsdorf apparently never got that message. Instead, he began to contemplate what action he would take should Germany surrender. He decided for the moment to wait until his fuel and other supplies were running low. With the exception of that subchaser off the coast of Maine, he had been unsuccessful in the seven months of this patrol, and he was determined not to go home without engaging the enemy and bringing honor to his boat, his crew, and himself. It might well be the last opportunity for action in the war.

The captain called the navigator into his quarters, asking him to bring the charts of Long Island, Block Island, and Buzzards Bay. This, it seemed, was the most likely area in which to intercept enemy shipping, as much of the traffic lane was well protected by Long Island Sound, with the Connecticut shoreline to the north and Long Island to the south. But ships coming out of the Sound were in open water between Montauk Point and Block Island, and then again between Block Island's east side and the island of Cuttyhunk on the west end of Buzzards Bay, which led into the Cape Cod Canal.

Froemsdorf decided that the Buzzards Bay end of the target area was too close to land for a submarine, as there were too many shoal areas that would hinder an emergency dive. He must have studied the charts at length, running out a battle plan in his mind, and finally making his decision. He drove his dividers into the chart. He might well have said, "Here is where we will fight. Here is where we will die."

As soon as the United States entered the war with Germany and Japan, all merchant ships had been nationalized into government service for the duration of the war. Therefore, Captain Charles Prior of the *Black Point* was master of a ship sailing under the flag of the U.S. Navy. In this capacity the vessel now carried a crew of five navy gunners for the five-inch cannon and single anti-aircraft gun. She also carried a complement of forty-one seamen and officers, and several of these were trained to help man the weapons.

Just after midnight on May 5, 1945, the *Black Point* headed east out of New York Harbor into Long Island Sound after completing an uneventful voyage from Newport News, Virginia, on her way toward Weymouth, Massachusetts. She was loaded with a cargo of 7,500 tons of soft coal and had left her coastal convoy at the approach to New York Harbor, as these waters were considered to be free of enemy submarines.

Still heading east while sailing inside Long Island Sound, the captain called for a minimum number of men on watch. Most everyone was in the galley drinking coffee and boisterously discussing the imminent declaration of peace. Germany had ceased fighting, and all that was needed was for military leaders to sign the initial peace documents. The war was over. The men could hardly contain themselves.

There would always be the never-ending battle with the seas of the North Atlantic, but these seamen were always willing to face this test, as seamen have for millennia. It was the threat of submarines that had sent a chill into the hearts of all aboard, but now the constant fear of a U-boat attack had faded from their lives. This cruise would be a breeze.

It was as the *Black Point* neared Montauk, at the easternmost end of Long Island, that Captain Prior called for all hands to be on deck and at their stations. Peace treaty or not, he was taking no chances as they entered the open sea. During the war, German submarines had been spotted in the narrow body of water between Montauk Point and Block Island, and they might well be in that area still. His biggest concern, though, would come as they passed the end of Block Island's north reef and its North End Buoy, better known by local fishermen as the Sandy Point Buoy. They would have to cross nearly thirty miles of open water before they would come under the protection of the islands making up the channel leading into Buzzards Bay. If there were subs lying in wait, they would attack here.

Looking at the chart, Prior could quickly see that the most likely spot would be in the first ten miles beyond the North End Buoy, as that area had the deepest water. And as he studied the chart, the first mate moved in to look over his shoulder and heard the captain mutter softly, "Here. This is where they will strike."

He looked up as he spoke, and the first mate assured him that the hostilities were over. The captain grunted and turned toward the helmsman.

"Keep your eyes open, son, and be ready to respond to my orders quickly."

"Yes, sir," replied the young man at the wheel.

"Mr. Thompson, instruct the O.D. to have all hands put on their life jackets and helmets, and to keep them on until we get inside Buzzards Bay."

"Yes, sir," snapped the first mate as he left the wheelhouse, little knowing that the captain may well have saved most of the crew with that order.

While the collier and U-boat were following the course that would shortly engage them in a life-or-death struggle, other activities were taking place less than four miles inshore—events that would play a large part in saving the biggest part of the crew of the *Black Point*.

Two new navy rescue boats were tied on either side of the dock at the lifeboat station in Galilee. The captain of each vessel was putting his crew through their paces in search-and-rescue techniques. These boats looked much like the famous PT boats that had earned recognition in the Pacific Theater of the war, and they had similar hull construction. But, rather than there being torpedoes and anti-aircraft guns in turrets on each side of the low-silhouette wheelhouse, their decks were clear except for the lifesaving gear needed for rescuing crews from downed aircraft.

As soon as the crews had checked all the equipment on deck, as well as the quarters and medical supplies below, they made certain that all items were accounted for and secured. Both boats were about to head out to sea. There the crews were to run through the drills they needed to become proficient in quickly hauling pilots aboard and getting them safely to more advanced medical care ashore.

All of the crewmen aboard these two rescue boats were billeted at the "Galilean," a large rooming house owned and operated by my aunt Della. During the war, the military would frequently house men—Seabees, rescue personnel, army engineers, etc.—in private establishments rather than build barracks on land that would have to have been confiscated from local people. This brought revenue into the area being served, made residents happy, and employed a number of civilians who opened their establishments to the armed forces. Aunt Della not only provided fine living quarters, but she also set a great table, serving the most delicious meals these military types had ever seen.

Of course, I used this familial connection to get myself invited aboard these sleek new rescue boats. I was fourteen at the time and was thrilled to be able to spend considerable time aboard the boats and with the crews during their off-hours at Aunt Della's. I had free rein aboard both boats while they were tied to the dock or were under way for any trips not related to rescue operations. Because of this privilege I was aboard one of the rescue boats on that glorious, warm May morning as the *Black Point* and *U-853* approached the point on the chart where their paths of destiny were to cross.

On the night of May 4, *U-853* surfaced just south of Block Island after running at periscope depth to check on the presence of any Allied ships in the vicinity. They waited at that depth until two fishing trawlers steamed off to the south'ard and out of sight. The expected radio signal from Germany to all boats in the wolf pack did not come through at the scheduled time. This caused great concern for Captain Froemsdorf, as he was without current instructions from Admiral Karl Doenitz.

This break in the schedule could have meant that the Allies had captured the submarine base or that Germany had surrendered, and Froemsdorf might well have had a brief moment of panic.

He would never learn that Allied bombers had destroyed the huge radio tower that Doenitz had used all during the war to communicate with the U-boat fleet throughout the Atlantic Theater of operations. Thus, each captain was on his own in deciding whether to surrender, fight to the death, or scuttle his boat. Captain Froemsdorf had apparently already made up his mind.

The command to dive was sounded throughout the boat. Froemsdorf was the last man to leave the tower. He took one final look at the beautiful array of stars in the heavens and then, turning to see the flashing red light of the lighthouse at Southeast Head on Block Island, he made a mental note that the decision he had made earlier was the correct one. A seaman waited until the captain was belowdecks, then turned the dogs (latches) on the outside hatch of the conning tower. It was now battened down just as the air tanks were blown, allowing the *U-853* to settle beneath the surface.

"Come left to five degrees, go to one-half speed," Froemsdorf said to his first officer. This command was passed on to the navigator and the helmsman. "Load all torpedo tubes, fore and aft." Again this order was passed along to the officer in each torpedo room. "Hold this depth, then maintain our position at 41.13 N and 71.27 W," he called out with a strong, firm voice. Again the first officer passed on these instructions.

Nothing more was said by the captain as he eased himself into his seat, which was secured to the deck within two steps of the periscope. *U-853* continued her slow journey.

"We have arrived at our destination, Captain," the first officer reported to Froemsdorf.

"Very good," he responded. "Please request all officers to meet me in the officers' mess, Karl," he said softly, as he eased himself off his chair and went forward toward the mess area, settling into his customary seat there. As the officers began to arrive he asked the chief petty officer to unlock the liquor locker. "Get a couple of bottles of schnapps for us. Karl, can you get glasses for all around?"

Both men quickly complied with the captain's request, then sat quietly, waiting for him to speak.

"Gentlemen, I'm sure you are aware of the recent developments at home. We were unable to receive our orders tonight, and I'm sure that means that we have lost the war. You are all aware that some of our comrades have scuttled their boats with all hands aboard, and I can understand their actions. But I do not want to end our mission that way. We're not sure how the U.S. Navy will react if we were to surface and offer to surrender. They might sink us or take us as prisoners of war. I don't know." Froemsdorf paused. "Gentlemen, fill those glasses and pass them around, please."

Nothing was said as the glasses were handed to each man. "To the Fatherland," said the captain.

All the officers repeated his toast and emptied their glasses.

"Please, fill the glasses again. Gentlemen, relax, please. I want to share my thoughts with you as to the decision I have made in this regard. I want your heartfelt reaction, not just what you think I want to hear. Understand?" he asked as he looked into the eyes of each officer present.

"Yes, sir," they responded in unison.

"Karl, will you please roll that chart out on the table? Gentlemen, lift your glasses, please," said the captain. With the chart spread out on the table, all watched as Froemsdorf stood, leaning over the table slightly and pointing with his pen to the exact bearings that he had selected a short time ago. "I plan to meet the enemy here," he said. "I also expect that we will die here, but then again, we might be lucky enough to reach the deep water to the south.

"There might even be a thermal layer of cold water nearby that would shield us from enemy sonar. I'm not hopeful of such an outcome, but there is always a chance: At any rate, I want to end our part in this war with a victory. We are on the offshore edge of a shipping lane. I plan to stay perfectly quiet on this bearing until we hear the sound of a merchant ship inshore of us. Then we strike!" declared the captain, as he pounded his fist on the outspread chart. "Gentlemen, are you with me?"

"We are, Captain," they answered as one.

"Good. Good. Thank you, men. You are loyal officers and I count it as a blessing that you are serving with me. We will be successful. Now, let us have one more drink and then those off watch should turn in. We want to be rested and as mentally sharp as we can be come daylight. Thank you again, gentlemen, for your loyal support." Standing, he said good-night to each man, shaking hands as they left.

Captain Froemsdorf left instructions that he was to be informed as soon as propeller sounds were heard; then he, too, turned in. He fell into a fitful sleep, awaiting his call.

The intercom buzzed twice before Captain Froemsdorf hit the switch. "Yes?"

"Good morning, Captain. Sonar just reported propeller sounds but not those of any large ships."

"Thank you, Karl. I'll be right along." As he spoke, Froemsdorf was rolling out of his bunk. He dashed some cold water on his face, brushed his thinning hair briefly, then set his cap on his head, adjusting it to the comfortable fit that he liked best.

"Good day, gentlemen," Froemsdorf said minutes later to the men in the control room.

A unified, "Good day, Captain" was the response.

"Sonar, how far off is that propeller activity?" he asked. He could feel excitement rising in his throat as he spoke.

"The sound seems to be affected by the shallow water, sir, but I would say no more than four miles and heading south at ten knots, Captain."

"Thank you, Karl. Are there any other ships in the vicinity?"

"No, sir. There were several smaller boats earlier this morning, probably fishermen heading out from Point Judith. Nothing else for this past hour, sir," responded Karl.

"Very well, gentlemen, we shall wait a while. Karl, please see that all hands are well fed and tell those off watch to get some rest. We want to be at our peak when the action begins."

"Yes, sir," replied the first officer, who quickly passed the orders along.

The men aboard *U-853* rested, read, and ate throughout the long day. Occasionally sonar picked up propeller noise, but it was always that of small craft, most likely fishermen going to and coming from the harbor at Galilee.

Finally the sonar technician spoke sharply. "Captain, I have loud propeller noise from a large ship."

"What direction is she heading and how far away?" asked the captain.

"It's difficult to be certain in this shallow water, captain, but I would say four or five miles, and she's heading east."

"Karl, record that the enemy was first heard at 1700 hours on May 5, 1945. We are preparing for action."

The first officer answered nervously, "Yes, sir. So recorded."

Captain Froemsdorf reached for the microphone on the PA system. "Attention. Attention, all hands. This is your captain speaking. We are very close to a shipping lane that runs right along the northeastern coast of the United States. A large merchant ship is about to cross our bow, and I have every intention of sinking it. This may well be the last opportunity to strike a victory for the Fatherland. I want every man to perform his duties perfectly. We are in very shallow water, so we may face some difficult times until we can reach greater depths offshore. Good luck and God bless you all." He then hung up the mike and turned to those around him, saying in as firm a voice as he could muster, "Gentlemen, let's go to war."

The *Black Point* came abreast of the North End Buoy precisely at 1700 hours as Captain Prior looked at his watch and said, "Mr. Thompson, sound general quarters."

"Yes, sir," responded the first mate, wondering what all the commotion was about.

A few minutes later, the captain requested some coffee from the officers' mess. Shortly thereafter the steward handed him a mug of hot, black coffee. Thanking him, Prior turned to ask the first mate the depth of water when he was lifted into the air, smashed against the overhead, and dropped to the deck. At precisely 1740 hours a single torpedo fired from

U-853 struck the collier aft of the magazine and astern of the wheelhouse, which sat amidships. Forty feet of the stern section was instantly torn away.

At 1755 hours, the *Black Point* capsized and sank in ninety-five feet of water just 3.2 nautical miles from Point Judith, Rhode Island. In the few moments between her being torpedoed and her capsizing, total chaos reigned aboard, and all the crew could do was abandon the ship as she was sinking beneath them.

Twelve men on the *Black Point* died in the first moments of the attack. Many more would have been killed had they not been standing by at general quarters with their life jackets on.

The torpedo had caused such destruction so quickly that the crew were not able to send out an SOS calling for assistance. Nearby, however was the SS *Kamen*, a Yugoslav freighter, and within two minutes of the attack on the *Black Point* they radioed word of the sinking.

Thus it was that the crews of the "crash boats" docked at Galilee were scrambled, and both headed out of the harbor of Galilee at flank speed, while their crews prepared to receive survivors. They were soon joined by Coast Guard "picket boats" from Block Island and Point Judith, along with fishing vessels that had been operating in the vicinity. The *Kamen* was on the scene, as well.

This quick action saved the lives of thirty-four *Black Point* crew members, all of whom were brought into Galilee and offloaded at the two docks at the lifeboat station. Ambulances then transported the injured to local hospitals.

I was waiting on the dock when these survivors were brought in, and I rejoiced with other onlookers to see so many men still alive after such a horrendous attack. Within an hour, the bodies of those who did not survive were also brought in. These, too, I saw, some badly burned from fuel oil that caught fire. Eleven seamen and one of the navy's armed guards were killed. This was no sight for a fourteen-year-old boy to witness, nor should anyone, of any age, have to see such devastation, especially when all observers knew that the war with Germany was over.

At 1742 hours, immediately after the explosion, the radio operator of the *Moberly*, a Coast Guard frigate traveling with two navy destroyer escorts, the *Amick* and the *Atherton*, had picked up the signal from the *Kamen*. These ships were thirty miles from the scene and arrived in the vicinity of the sinking at 1930 hours. Taking stations some 3,000 feet apart, they began their search.

For the remainder of the evening, a series of attacks on the *U-853* ensued. Each time the vessels believed they had dealt a mortal blow to the German sub, their sonar would reveal its movements as it attempted to escape. The navy used hedgehogs—rocket-launched projectiles—and depth charges. The battle was a costly one. Shortly after midnight on May 6, the *Moberly* and the *Atherton* were both damaged when they failed to avoid the explosions of their own depth charges. Eventually, as the evening wore on, the attacks were halted until 0530 the following morning, when the sun began to rise on what would be the final day in the life of *U-853*.

Two blimps, the *K-16* and *K-58* from Lakehurst, New Jersey, joined the attack with the arrival of daylight. They were directed to assist in locating and identifying oil slicks, and to help mark the location of the submarine with smoke flares and dye markers. The U-boat was believed to be heavily damaged, and it appeared to be bleeding large amounts of air and oil.

The *K-16* blimp dropped a sonar buoy on a spot where oil was still rising to the surface. The sonar operators in both blimps then heard the sounds of metallic hammering coming from the submarine. About ten minutes later, they heard a long, shrill shriek and the hammering stopped.

The blimps then made a series of attacks on this spot using the 7.2-inch rocket bombs, and at 1045 hours, *U-853* was declared sunk and on the bottom, 7.7 miles east of Block Island. The navy vessels headed for port with brooms at the masthead, the symbol for a clean sweep.

On May 6, 1945, navy divers from the vessel USS *Penguin* dove on *U-853* in order to recover records, ship's logs, codebooks, and anything else of importance. They were successful in these efforts and in confirming that the sub's captain and crew had suffered the same fate as that of their dead leader, Adolf Hitler.

Today, *U-853* sits in a hundred and thirty feet of water off Block Island, upright and intact on a sand bottom. She has the sad distinction of being one of the horrors of war after the conflict had officially ended.

And her victim, the *Black Point,* is in the annals of naval warfare as the last American merchant ship to be sunk by a German submarine—and that on the day that Germany surrendered.

Every voice that I heard that day as the collier's wounded and dead were brought ashore echoed the same mournful refrain: "Why?" . . . "Why now?" . . . "It was over!" bemoaning the ugliness, the utter futility, and the total waste of war.

It calls to mind the observation of a wise old prophet when he stated, "All this I have seen, and there was an applying of my heart to every work that has been done under the sun, during the time that man has dominated man to his injury." (Ecclesiastes 8:9)

THE *JANE LORRAINE* AND THE DECEMBER SURPRISE

Fair fishing days in winter are rare. When a weather report indicates that the next sunrise will precede such a day, every captain in Galilee is up early and ready to go. So it was that clear, cold morning in December 1947.

Leon "Buddy" Champlin arose well before the sun's rays broke over the eastern rim. Boarding the *Jane Lorraine*, he started the engine and put into motion a day like never before. As Buddy checked the electronics and turned on the deck lights and running lights, he gave his brother Ken the nod to lower the net to the deck prior to heading offshore. Buddy's brother, John, and Cliff Whaley quickly guided the twine, pulling the chain hung along its mouth snug alongside the starboard rail. The head-rope, with metal cans attached every few feet to lift the top of the net off the bottom, was lowered and placed inboard just a bit. Then the body of twine was folded carefully, allowing it to be set overboard in one smooth motion. The last part of the net—the "cod end," or bag—was placed carefully on the bulk of the twine, and one man tied the series of knots in the bag's "puckering string." This arrangement would keep all the fish securely inside the trawl as it was being towed and, later, hauled onto the deck, but it would quickly yield them when the series of knots was tripped.

Buddy had served for several years in the Merchant Marine during World War II and had been exposed to enough danger to last any man a lifetime. About a year after his return home he laid out his plans for an eastern-rig dragger to Cliff Whaley, who was at the time a local ship-builder of considerable renown. Cliff had built some pretty little draggers, one of which his brother, Joe, owned—the *Virginia Marise*. In fact, Joe fished that boat right up to the time of his death.

Buddy's plan called for a hull of forty-eight feet in length, thirteen feet in width, and nine feet in depth. She was perhaps the deepest dragger of that size in Galilee, and her draft, along with a hull that was extremely seakindly, made her a very comfortable boat. She was built to last a lifetime or beyond—at least, that was Buddy's plan.

The members of the Champlin family were all seafaring men, and most were fishermen. Buddy's father, Leon Sr. (better known as "Brownie"), was a lobsterman extraordinaire. I fished with him during the first year of my marriage. Early in the lobster season we set a number of pots (traps in Maine) close to the beach and rocks along the east side of Point Judith and up near Narragansett. As we were hauling each pot, Leon would invariably tell me exactly what would be in it: an "egger" (a female bearing eggs), a one-clawed "bug," an undersized "short," and so on. He never failed to guess correctly, and had I not been present I would not have believed it. Later in the season we would set ten-pot trawls offshore, most of the time around Cox's Ledge.

Buddy, too, had an instinct about where to catch fish and when they would be there, and it was not long before he became a highliner in Galilee. He could and would fill his nets where no one else would think of going.

His first year of fishing had been good. As he developed his skills, they would become even better; he also had the kind of boat that would enhance those skills. Buddy was looking forward to the upcoming New Year, just a few weeks away.

On this cold December morning, as the *Jane Lorraine* left the East Gap of the breakwater and headed about southeast, as I recall, she had a two-hour steam before the crew would "set in" at Deep Hole. Buddy liked fishing there. It was tricky holding the edge, and what with the rocks, wrecks, and other obstacles, the helmsman had to stay on his toes if he hoped to get in a full tow without hanging up on the bottom. Yet when he was sufficiently diligent, he usually got a good haul of flounder, whiting, and a variety of other marketable fish. A full day of such tows would yield a very profitable catch.

This was not to be a full day of fishing, however—not today nor a lot of tomorrows.

When the depth sounder and the loran showed Buddy the edge of Deep Hole he wanted, he slowed the engine and began turning to put the starboard side into the wind. The crew came scrambling out of the fo'c'sle, with Ken still holding a mug of coffee as he headed toward the winch. He set the coffee down between his feet, threw the winch into gear, and in one smooth motion lifted the drag doors off the deck so the other men could swing them overboard. With a quick turn the crew began to heave the net over the side, beginning with the cod end and then the body of the net, which drifted off as the wind blew the boat clear of the trailing twine. Next, the mouth of the net, with its headrope and footrope, went into the water. Buddy then turned the *Jane Lorraine* to starboard and increased speed until he had the engine at the right rpm. He was heading in just the right direction when he leaned out the window and told Ken, "Let 'im go."

Down went the doors as Ken held just the right strain on them to keep them spreading as they settled into the water. Buddy watched the tow wires until the hundred-fathom marker came into sight; then he slowed the boat long enough for Ken to set the brakes on the winch and for John to hook up the wires at the gallows frame aft along the starboard rail. At that point, he resumed the speed that he used towing in the "Hole."

The crew went below to get some breakfast. When he had finished eating, Ken came up into the wheelhouse to relieve Buddy, allowing the skipper to go below for a quick meal. You did not dally over food when you fished this area. Shortly after Buddy had returned to the helm and the crew had prepared the deck for the first tow, the boat was headed into the offshore leg of the Hole, and it was time to haul back. Buddy slowed the *Jane Lorraine* enough to allow his crew to unhook the tow wires and for Ken to engage the winch that would haul the net in.

The doors broke water, were brought forward to the side boom, and were separated from the running gear. Next, the wire-cable "legs" leading to the wings of the net were wound onto the winch. In short order the trawl was brought up to the side of the boat as Buddy kept steaming slowly in a circle until he again had the wind on his starboard side. He threw the engine out of gear, and the men on deck went to work putting a strap around the body of the net, beginning to hoist it overhead, and shaking the twine to move errant fish down into the cod end, which was still floating at the surface of the water.

Then a strap was slipped around the top end of the bag, the big hook attached to the main falls (the ropes of the lifting tackle aloft) was inserted into the strap, and Buddy began to hoist the cod end aboard. As the bag swung inboard and over the rail, he dropped it on deck just long enough to stop its motion, then raised it quickly and just high enough to let John trip the rope (the puckering string) beneath the struggling fish, allowing the catch to flow out onto the deck. A quick check assured the crew of a good day, as several bushels of yellowtail flounder were mixed in with a good supply of other marketable fish.

The net was reset and after a few more tows, all bountiful, Buddy was looking forward to another tow or two and hoping the weather would hold. At this time of the year it didn't take long for things to get sloppy, and working here in the Hole conditions could get dicey in short order.

As the sun fell low on the horizon, the last tow of the day was about to come aboard. John let the tow wires go, and Ken put the winch in motion and began hauling the gear in.

Just as the drag doors broke water, Bud noticed that the net seemed heavier than usual, even for a good tow. It was hanging lower in the water than he wanted to see it, so he made a slow turn to starboard. The doors were up all the way, and there were about one hundred feet of steel wire legs were trailing behind the boat, as was the entire length of the net, which was dead astern and coming closer to the surface. Buddy was just about to back down while those legs were hauled in, but he never got that far.

The explosion was so great that it literally blew the seawater out from under the stern of the *Jane Lorraine*. It lifted the after section of the boat into the air, bending the planks and keel upward from their normal position.

Just before disaster struck, Ken had stepped up onto the massive, two-piece wooden hatch covers over the fish hold in an effort to discern why the net was so heavy. With the explosion, the covers flew into the air, Ken dropped a dozen feet to the concrete floor in the fish hold, and the covers came slamming down on top of him. Buddy, who had been leaning out the wheelhouse window to talk to Ken, was slammed upward into the ceiling. The electronics, secured by bolts to the wheelhouse frame, were torn free as though there had been no fastenings whatsoever. The radio direction finder (RDF) that had a circular antenna used to ascertain the vessel's position, was smashed into the overhead, crushing the antenna nearly flat when every bolt holding it securely was sheared off by the force of the explosion.

In the engine room—where the diesel was secured by sixteen ⅝-inch-diameter, case-hardened steel bolts—the engine was blown onto its side. The bolts appeared as though cut by an acetylene torch. In addition, the torque of the jammed shaft and propeller had stalled the engine, causing the loss of needed bilge pumps, and inward water pressure from the explosion had blown the two-inch seacock out of the hull, allowing a tremendous flow of seawater to flood the engine room.

Although stunned by the explosion and suffering from confusion that quickly dissipated, no crewmen were injured, and that was of prime importance to Buddy. He immediately sent out a "Mayday" over the frequency used by fishermen, hoping against hope that the radio was not damaged beyond use. No sooner had the call gone out than Howard Whaley on the *Harold and Bruce* responded that he was on the way. The *Tip Top,* owned by George Gross, also acknowledged that he was going to help. Both boats were from Point Judith and were also fishing Deep Hole, but were up at the north end when they heard and felt the explosion. The impact had been so strong that each captain thought his boat had been the one struck by the explosion. When Howard and George realized that they had suffered no damage, they quickly knew that one of the small fleet fishing the Hole would be needing immediate assistance.

Back on the *Jane Lorraine,* Buddy had Ken, John, and Cliff use the block and falls to lift the doors upward and swing them on deck. Doing so would prevent them from being in the way when Ken and John received towlines from the two boats coming to assist. It would also ensure that the doors wouldn't drag in the sea while the *Jane Lorraine* was being towed, causing the dragger to yaw to starboard. The remains of the net were then hauled in and stowed on deck for the same reasons.

Next, Buddy had the crew pull some tow wire off the winch drums to make up a strap, or yoke, to which towlines from the boats that were on the way could quickly be secured, as every minute counted in getting the *Jane Lorraine* under way and headed back toward Point Judith. This cable would be run out of the hawsehole and then looped around and secured to the towing bitt forward. In an effort to prevent any trouble if the bitt was not sturdy enough, another cable would be wrapped around the base of the mast, as it was stepped into the keel and would be amply strong for securing the tow wire from the two boats on their way.

Before this preparatory work began, and before Ken was certain that help was on the way, he began firing distress flares, one after another. Dennis Sidel of the *Carol and Dennis* out of Stonington, Connecticut, a known humorist in the fleet, called out on the radio, "You've got so many flares going up it looks like the Fourth of July. What's going on over there?"

Howard Whaley filled him in on what had taken place and told Dennis that they had everything under control. Dennis wished them Godspeed and offered to assist later if needed.

One other preparation made by Buddy was to "lock" the rudder of the *Jane Lorraine* dead center, ensuring that the tow would be as straight as possible. Fortunately, the helm turned—with great difficulty, but it turned. Simultaneously, Ken and John were down in the lazarette (small, watertight) compartment at the stern, trying to turn the quadrant that was bolted tightly to the shaft of the rudder. The combined strength of the three men finally got the rudder into the needed position. The two down below scrambled out of the engine room fast, as water was rising at a steady clip with the two-inch seacock gone. In the wheelhouse, Buddy put beckets on both sides of the helm and hoped that they would hold for the duration of the trip home. Boy, that was a nice-sounding word, right about now.

Things were coming together well under the circumstances. *If only the watertight bulkheads hold and keep the seawater within the engine room,* thought Buddy. A close inspection of the entire boat failed to turn up any other appreciable flooding. Still, a freshening sou'west breeze and increasing ground swell was adding to the difficulty of the preparations aboard the *Jane Lorraine*. Water within the hull was sloshing so much it was becoming harder to get any repairs done belowdecks.

As all of this was going on, Jack Westcott brought the *Joyce Ann* alongside to offer any assistance that might be needed. Buddy told him that Howard Whaley and George Gross were coming to tow him in to the Point and thanked him for his offer. Just as Jack was about to move away from the *Jane Lorraine,* his brother Chet asked him to bring the *Joyce Ann* in a little closer; he wanted to jump on board to help in any way that he could.

Jack knew that with the freshening wind and seas, getting too close could be dangerous, yet he could see the grim determination on Chet's face so he said, "Wait until I back away and come up on the leeward side of her. That will give you a few seconds of calm before you jump."

As the *Joyce Ann* maneuvered into position on Buddy's port side, all hands on both boats could see the churning ocean between them. If Chet missed his mark or lost his footing upon landing, he would surely be crushed between the two hulls.

Jack came in slowly and about as close as he dared, what with both boats heaving in the increasing seas. He called out to Chet to get up on the caprail of the *Joyce Ann* and to wait for his signal. Chet quickly grabbed the side stay that ran down from the mast to the after gallows frame, and this put him within a few feet of Jack, who was hanging out of the wheelhouse window. "All right, get ready! Steady now . . . Wait for this next sea . . . Go!" he yelled, just when the two hulls paused momentarily as a sea slid between them.

Chet put all of his strength and energy into that dive toward the *Jane Lorraine,* and as soon as his feet hit the caprail of the bulwark on Buddy's boat, both Ken and John grabbed him and pulled him inboard, all of them landing on deck in one big pile. It was not Chet's most graceful moment, but his efforts were successful.

As soon as Jack saw that his brother was aboard, he threw the engine of the *Joyce Ann* into reverse and gunned the throttle, putting as much space between the two boats as possible. He then fell in astern of Buddy and stood ready to help if needed.

Quickly untangling themselves, Chet, Ken, and John raced toward the engine room to see what might be done to stop the flooding. They found that the water had risen to the overhead, and there was so much of a surge as it sloshed from side to side that they couldn't get to the seacock. They had hoped to put some kind of patch over the area that once was the through-hull entry point, and perhaps to jam a timber between the tipped-over engine and the patch to at least slow down the flow. Unable to pursue that course, they fell to taking turns on the deck pump. At least that would suck out a five-inch stream of water as long as someone was manning the handle. Chet offered to take on this task while John, Ken, and Cliff prepared for the two boats coming to tow them in.

The *Harold and Bruce* was the first to arrive, and her crew threw a light line to the *Jane Lorraine* so Buddy's gang could begin hauling in the tow wire, securing it to the wire yoke they had prepared for the job. By the time that was completed, the *Tip Top* was alongside and they, too, were quick to get their towline aboard. Next, both vessels ran out several hundred fathoms of wire until they both had released the same length for the tow. Then they gave a signal that they would begin to slowly take up the slack until a proper strain was equally distributed between both draggers.

As soon as the *Harold and Bruce* and the *Tip Top* were under way, towing the *Jane Lorraine* toward home, Howard Whaley called the Coast Guardsmen at the station in Point Judith and advised them of the situation. Of course, the Coast Guard was quick to offer assistance.

Howard assured them that as long as the three draggers continued to make the headway that they had managed so far, it would be more expedient to continue as they were. He said that it would be good, though, if the Coast Guard would stand by in the event they were needed.

Soon forward speed lifted the crippled boat higher in the water as it flowed ever faster beneath the hull. It was imperative that there be no break in the headway that was keeping the *Jane Lorraine* afloat. Progress was better than anyone aboard the three boats had expected. The breakwater was soon in sight, and Howard called George aboard the *Tip Top,* with Buddy listening in on his stove-up radio. The two skippers agreed that once they were inside the breakwater, Howard would release his towline, then swing back and come up alongside the *Jane Lorraine* on the starboard side. Buddy, he knew, would want to put his port side to the dock, so Howard planned to ease in, planting with his bow just about a third of the way forward of Buddy's transom.

The crews of the two boats would then link them with bow lines, bow springs, stern lines, and stern springs, secured fore and aft. This would put two-thirds of the *Harold and Bruce* out behind Buddy, allowing Howard to maneuver both boats in the close quarters inside the breachway and around the docks, putting the port side of the *Jane Lorraine* against the pilings. This plan went like clockwork, and the necessary changes were soon made. When both boats were secure to each other, the

Tip Top released her towline, and the *Harold and Bruce* took full control, bringing Buddy in as slick as could be.

Some seamen from the Coast Guard station were on the dock, and they quickly jumped aboard the *Jane Lorraine* with some pumps to keep her afloat until arrangements could be made to haul her out at the shipyard the next day.

At daylight, Howard brought his boat in as he had the day before, backed both draggers out into the channel, and maneuvered Buddy into the cradle on the railway at Captain Knight's shipyard up in Snug Harbor. Then he backed away, allowing the *Jane Lorraine* to be hauled out and prepared for the extensive repairs that were to last many months.

Everyone was amazed that the *Jane Lorraine* had taken such a beating yet continued to remain afloat. Even Cliff Whaley was pleasantly surprised at the strength of the hull he had designed and built. Three planks running from the stern well forward, toward amidships, had sprung from the explosion but had not come loose enough to cause flooding. The section of the hull from about the vessel's transom to a point forward of the wheelhouse had been lifted right into the air, but it had settled down again, causing the damage to the seacock, engine, and electronics but creating no opening for seawater to flood other sections of the dragger. Yes, she was a well-built boat, one that Cliff could be proud of.

In a convoluted way, this experience proved Buddy Champlin to be a fortunate fisherman. Had the *Jane Lorraine* been an older boat, or one not built so soundly, the torpedo or depth charge or whatever it was that had exploded, would without doubt have shattered her hull. Had the munitions trapped in the net exploded under her keel instead of a hundred or more feet behind the boat, it is a certainty that the *Jane Lorraine* would have disintegrated, and all hands would have been lost.

And what if the device had been brought right up to the side of the boat with all hands looking over into the water to see just what was giving them so much trouble? *Boom!* Good-bye crew. In addition, Buddy had fished the Hole many times when no other vessels were in the vicinity. Had that been the case on this day, the results of the explosion would have been disastrous.

Yes, Buddy Champlin's December surprise, although costly, really turned out to be a December gift. He had started this fishing trip with the prospect of a successful year ahead of him. Because of a wonderful series of circumstances that saved his own life as well as the lives of three men dear to him, Buddy could indeed look forward to a happy, productive career and a long life.

As this is written, he is still fishing out of Galilee on the *Jane Lorraine*.

CAPTAIN'S LOG

Section III
The Final Years

MISTRESS OF THE SEAS

As a small boy raised at the edge of the sea, I tried to absorb the terminology used daily by the fishermen around me as they carried out various tasks aboard their vessels and ashore.

I was fascinated with the names and meanings of such things as "sea anchor"—a canvas cone that floated when thrown overboard while most anchors sank instantly to the ocean bottom; a "monkey's fist"—a special knot that did not look at all like the hand of a primate; a "turtle back"—an extension aft of the wheelhouse—and a "whaleback," or raised deck, running from the bow aft to the mainmast. Both were common aboard eastern-rig draggers, but neither looked like a turtle or a whale.

I gradually learned this unique language and adopted it as my very own. Yet there was one thing I never really understood until I grew to manhood and became the master and owner of my own vessel: Why do men—all men—refer to ships as being of the feminine gender?

When talking about his fishing boat, the *Olive*, my grandfather would soften his voice and speak in a gentle tone he never used when referring to his wife or daughters. There was no question that he had a deep and abiding love for his family, yet he reserved his most intimate tone for his boat.

One day we were hauling the *Olive*, gently sliding her onto the half-submerged cradle at the boatyard. Secured by cables that were wound around the drums of a winch, this frame sat on railroad tracks that slanted down into the water. Any vessel to be hauled out for repair or painting was eased into this cradle, which then was gently pulled up out of the water onto land. This all took place at Captain Hanson's shipyard, up at the head of Salt Pond.

Because I was smaller and more limber than any of the crewmen, my job was to climb under the boat and begin to quickly scrape her bottom of the barnacles and sea grass that had been accumulating there all summer. It was imperative that this process begin while the hull was still wet, as the marine growth would bond to the wooden planks like concrete if it was allowed to dry. This was a horrible job. Not only did the growth drop down on me in a soggy mass, but it also went into my eyes, down my shirt, and into my boots. And as I scraped, the "red lead" antifouling paint would drop onto my skin and begin to sting like a wicked sunburn—I guess from the chemicals in the paint. I was always determined to complete the job as quickly as possible because my grandfather expected that of me, and I never wanted to disappoint him. But that didn't make the task any easier.

On this particular day, when the job was complete Gramp took a long, careful look and then said, "There now, you've made her happy. Good job."

As he turned to go, I asked him that age-old question: "Gramp, why do you talk as though the *Olive* is a girl?"

"Well, son, she is a lady. Didn't she provide you with a job? Didn't you feel safe aboard her? She carried you well when it got rough outside the breakwater, didn't she? When we were heading in the other day I saw you resting up for'ard; you even fell asleep. Didn't you feel the way you do when your grandmother holds you?"

"Yes," I answered, "but I still don't understand."

"Only a warm, loving woman would treat you that way. The *Olive* cares for us, all of us, so we have to take real good care of her." Then he turned away to speak with Captain Hanson.

I didn't really understand, but I knew that in time I probably would, and I did.

Boats and ships are spoken of in feminine terms because they truly are the mistresses of the men who own them. Let's consider a few reasons why I say that.

Webster's Ninth New Collegiate Dictionary states that "mistress" is derived from the French word, ***maître*** (master) and defines the term as: 1) "A woman who has power, authority, or ownership." 2) "Something personified as female that rules or directs." A skipper would therefore show deference to his boat when it came time to make a decision as to whom was served first. Let's consider a few situations that might help you to see why I believe every boat or ship is really a "mistress."

(a) A storm is working up the coast; it turns into a hurricane and might hit the captain's home port. What does he decide? He says to his wife, "Take the kids and go to your mother's until after the storm."

Her response is, "Where are you going?" She already knows, but she asks anyway. "I want to be sure she's all right. I may have to take her out to sea. I don't want anything to happen to her."

(b) "Honey, let's take the kids on a trip to Disneyland," suggests a fisherman's wife.

"I'd love to, but I need the money I've saved up to overhaul her engine," the skipper responds.

(c) A captain's spouse says, "Let's get the house painted. It's really looking shabby, don't you think?"

His reply: "I'd love to, but she needs a new loran and I have to repair her wheelhouse." Notice that the wife doesn't even ask who "she" is; she well knows.

(d) The crew was at sea for eight days. They were in some bad weather, but they finally got a trip (a successful catch of fish).

The skipper climbs into bed with his wife on his first night home, but in the middle of the night she rolls over and reaches for her husband. He's not there. Is she concerned or worried? Not at all. She knows where he is; he's down on that @%#^&** boat. He was worried about keeping "her" pumped out, and he wanted to check the fire in the fo'c'sle stove.

He goes into the wheelhouse with a cup of coffee in his hand, opens one of the windows, and looks over his mistress. He runs his eyes up and down her deck. He gazes fondly at her rigging. He runs his hands over the wheelhouse window frame. Then he just stares. After a while he heads back home. No questions will be asked, and his wife certainly won't challenge his decision as to whom he will pay the most attention. The wife well knows that the mistress will win out.

I learned the true power of the *maîtresse*, the mistress, that siren of the sea, when I purchased my first dragger, the *Roberta Dee*. That's when the infatuation began. Although I was excited, I was also fearful, and I think that dampened whatever romantic feelings might have been dwelling within me. I had my first fishing vessel, my first mortgage (yes, my first mortgage was for my boat, not a house), and my first crew—all married and blessed with children. Not only did I have to catch enough fish to pay my own debts and care for my wife and children, but I also had three other family units to provide for. At twenty-two years of age, that appeared to be a big nut to crack, so my energies were directed toward these responsibilities, not a shipboard romance.

It was not long, however, before the siren call was heard and the passions began to emerge. Stonington, Maine, was where it all began. I had lost the *Roberta Dee* in a storm, all hands had survived, and the bank had been paid in full. In fact, the loan officer was so pleased with my fishing record that he said the bank would go up to 50 percent on any vessel I wanted. I had a new lease on life, a new beginning, and I was anxious to get under way.

After several short trips down east, looking for just the right boat and not finding anything that satisfied me, I made one more try. My wife and I arrived in Deer Isle and hit every little harbor on that handsome island. We finally arrived in Stonington, a beautiful fishing port on the southernmost tip of the island. It was a lovely sunny day and all the local draggers and lobster boats were at sea. There was one more dock to check, and as we drove up to the bulkhead and parked the car, the angels began to sing. Music of the heavens exploded over my head. The power of the *maîtresse* had struck. Lying at the last dock in the harbor was the *Dorothy & Betty II*—sixty-five feet of wooden beauty glowing in the rays of sun that were beaming down, as pretty as could be. I just stood there looking. My wife was saying something about lunch, but it did not penetrate my thoughts. "I found her! At last, I found her," I exclaimed. I headed for the dock, and my wife knew that lunch was off for a while. She could see that look in my eyes, and she'd had enough experience to know that all I was listening to was the call of the siren.

I tried to be cool as I approached that boat. All hands were busy loading gear aboard.

The vessel had just been hauled out and gone over from stem to stern, topside to belowdecks, fo'c'sle to engine room. I had never seen a more beautiful dragger. When I went aboard, I found that the woodwork in the fo'c'sle and wheelhouse was brilliant, with a hard, clear varnish. You could lie atop the engine in a white shirt without getting it dirty. A brand-new RCA radar, which I hadn't seen on a dragger before, had recently been installed. I was awestruck, not only with the radar but with all of the electronics and the rest of the wheelhouse. The *Dorothy & Betty II* was like a yacht, if you'll pardon the expression.

After I had regained my composure, I began a discussion with the owner/captain, Elmer Gross. He was small of stature with a weatherworn face and a smile that warmed my heart. We talked about the areas that he fished, the quantity of fish on those grounds, and the like. I was saying to myself, "You better just jump into it, boy, the day is dragging on and this man is busy." So I just let it out. The call of the *Dorothy & Betty II* was casting its spell. I calmly asked Elmer, "If you were to sell this boat, what would you ask?"

He thought for a moment, then gave me a figure that he undoubtedly thought would cause me to faint.

I said hopefully and without a tremor, "I'll buy her then. I'll be back by Friday with the money." And with that I put out my hand and Elmer, with glazed eyes, shook it.

"See you Friday, then," I said as I quickly climbed up on the dock. Without looking back I caught my wife by the arm and hustled her into the car. I turned around and headed up the small road leading off the island. As I drove, I explained to my wife that I had just bought the most beautiful boat I had ever seen, and that we were coming back Friday with the funds. She sat quietly next to me. She had recognized before I had that she was now going to be competing with the siren that all fishermen's wives had talked about since we had married.

When we drove into the yard in Wakefield, Rhode Island, I could hear the phone ringing. I ran into the house to answer it and found Elmer Gross on the other end.

The first words out of his mouth were, "I don't want to sell that boat. I don't know what got into me to even talk about it. I know that I shook hands on it, and I will honor that handshake if you insist, but I do not want to sell the *Dorothy & Betty II*." He, too, had been hearing the siren call, but it was too late.

I said, "I will hold you to it, Elmer. I'll be up with the money."

There was a moment of silence on the other end of the line, and I could hear the despair in his voice as he said, "Okay, I'll go through with it." And he hung up.

I felt really bad about pressing Elmer like that. He was a wonderful man, but the siren call I was hearing was stronger than my conscience. From the expression on my face, my wife understood what had transpired and knew that I had won out.

Now I had to begin the work necessary to accomplish my goal. To satisfy the needs of the siren that now possessed me, I was willing to go to extremes in the search for the funds to meet my commitment to Captain Gross. I had to find thirty-five thousand dollars before Friday next, and I was far short of that.

I lived in a small town where everyone knew my grandfather, and not only knew me, but also were aware of the fishing record I had chalked up with my first boat. I literally went from store to store, talking with each business owner and telling my story. I offered to repay all loans within the year, and I was willing to give them 8 percent simple interest. This was when 2 percent was the going rate for bank savings and well before the concept of certificates of deposit.

The week went by quickly. I only had one more day before I was supposed to head back to Maine, and I was five thousand dollars short. I approached the local Chevrolet dealership, repeating my story and offer. The owner, John Yemma, listened quietly, then turned to his brother and said while puffing on his cigar, "Eagle, give him the money." Eagle gasped and choked a bit, then agreed to come up with that $5,000.

With the needed funds in hand, I headed back to Stonington, Maine. I was going to respond to the whisper that had been sounding in my ears since I had first laid eyes on the *Dorothy & Betty II*. She would soon be mine.

After the business was transacted and ship's papers were transferred, my crew and I headed toward the boat. With an unnatural feeling of pride and responsibility, we cast the lines off the dock, turned around in the harbor, and headed to sea, soon to be in Point Judith.

Steaming into the harbor at Galilee was probably the proudest moment in my fishing career. Of course, the day of my marriage and the births of my two children were all exceptional memories. Still, as I said at the outset, the bond between a captain and his boat, his mistress, was beyond natural emotion. The *Dorothy & Betty II* was beautiful. She was very, very beautiful. And she was all mine—or, rather, she now owned me.

MANY ARE BROUGHT LOW

It sounds simple enough: "Stay in the fresh air, eat lightly, avoid fried or spicy foods; if troubled, rest lying down with your head low, in a comfortable, well-ventilated place." This is from the *Encyclopedia & Dictionary of Medicine & Nursing*, 1972 edition.

At what is this advice directed? Seasickness. Discomfort caused by the motion of a boat under way. The symptoms are nausea and vomiting, dizziness, headache, pallor, and cold perspiration.

I have been fortunate in that I have never been bothered by the slight rolling and gentle lifting of a fishing vessel at sea, combined perhaps with fumes from her hot engine and the pungent smell of rotting fish overlooked during cleanup. Are you still with me? Some people are brought low simply by reading a sentence such as the previous one.

In fact, the statement that seasickness is brought about by the motion of a boat under way is a little misleading. As a fisherman out of Point Judith who often took landlubbers aboard for a day, I saw great numbers of people brought low as soon as they walked out onto the dock and merely *observed* a vessel that was tied up and shifting ever so slightly in the tide.

On the other hand, there were those who were unaffected by the motion of the boat as it sailed out of the harbor, encountering the usual heave of ocean swells. These people usually began to taunt the other, less fortunate folks who had been quietly dying since leaving the dock in Snug Harbor.

But as soon as the *Olive* swung around and stopped alongside the trap barrels, she began a slow easy roll, rising up and down in the ground swell. Right away, the complexion of these "easy riders" began to change to a mint green with a streak or two of creamy saffron. Then a calmness of personality set in, and their boasting soon ceased. Oftentimes, if we were still out hauling traps into the afternoon, and the typical summer sou'westerly breeze picked up, it would not be long before all the visitors aboard were hanging their heads over the rail, hoping for a quick and easy death.

Of course, that all changed once we got back inside the breakwater and started steaming up into Salt Pond. When those weary survivors began to smell land, and the horizon once again stabilized, the natural color came back to their faces. It was amazing to see them restored to a measure of health, become somewhat joyful about life, and even claim to have enjoyed their experience at sea.

Over the years, as I worked on a variety of boats and with a number of different shipmates, I witnessed a range of strange reactions to *mal de mer,* the malaise of the sea.

One very competent skipper I worked for would come down to the dock early in the morning and tell me where he wanted to fish that day. He would then go below, lie alongside the engine, and try to go to sleep. The full responsibility of preparing the boat for that day's fishing fell upon me. The other crewmen and I would lower the nets to the deck, let go the lines from the dock, and steam out of the harbor toward the chosen fishing ground. When we arrived, and while it was still dark, we would set the net and drag doors, and begin towing.

Of course, I well knew how to hold such ground, and would continue to do so for an hour or more. If the sky was still dark when it was time to haul back, then the other crewmen and I would do so, and if the tow was profitable, we would set the net again. By the time we needed to haul the second tow, daylight had most likely arrived. I would then go below to wake the captain, who would come topside and take charge. Once he was up, the wind could blow a living gale and he was not bothered. How could this be? His particular form of seasickness required him to be able to see the horizon. Nothing bothered him, in fact, until we lost the horizon again in darkness; then down below he would go.

I well remember the time that a gale of wind came up from the southeast. We were in a real snarl of weather as we neared Point Judith. I went below, told the skipper of our problem, and suggested that he come up into the wheelhouse as we were approaching the breakwater and the seas were smashing over the wall. He just hunkered down and told me to take her in. Once I got us up to the dock, he came on deck and helped us secure the boat for the night.

One cook that we had on the *Dorothy & Betty II* would get deathly sick as soon as we left the dock and would remain so for the first twenty-four hours of the trip. Once that period went by, however, he straightened right out and was able to work belowdecks in a gale of wind. I was sure happy to see that, as he was the best cook that I had ever shipped with. Because of his skill, all aboard were willing to pick up the slack for his spell of disability.

Several times during the summer months I used my boat, the *Dorothy & Betty II*, for domestic missionary work on the islands of Nantucket and Martha's Vineyard. I would invite several men to make the trip with me, including one fellow from Providence. He just loved to come along, but he had one problem: as soon as he stepped aboard the boat, and even before I let the lines go, he became violently sick. Wobbling badly, he would climb down into the fo'c'sle and up into one of the upper bunks—mine, in fact—dragging along a five-gallon bucket previously used for lube oil. He would hug that bucket close to his chest and roll slightly to the right, wedging it up against the side of the bunk.

Again and again he would blow his guts into that bucket. One of the other men would take it up on deck, throw the contents overboard, wash it out with the deck hose, and bring it back to Bill, still in the bunk. This would go on all through the night until we finally reached Nantucket. Once we were docked, Bill would slowly and very weakly climb up on deck. Once on the dock he quickly came around, would take in some nourishment, and be ready for a day's work in the ministry.

I frequently told him that he did not have to go through that torture each time and should just stay home. He pleaded with me to let him go with us, and so he did. It must have been the Holy Spirit that gave him strength.

One time we were getting ready to sail on the *Roberta Dee,* my first dragger, planning to go to the south'ard of Muskeget Channel, between Martha's Vineyard and Nantucket. We would be fishing offshore in thirty fathoms of water, working on butterfish and some red hake and whiting. I didn't expect to be gone more than a couple of days. We had just iced up the hold and topped off our fuel tanks, and we were about to shut everything down for the night. I planned to leave the Point around one in the morning.

While we were cleaning up the deck, a middle-aged man dressed all neat and trim, and looking every inch a military type, was standing on the dock, watching us as we finished.

He asked if he could come aboard and I said that he could. He asked me if I had any opening for a crewman. I assured him that we had a full complement, and he said that he wanted to go out on one of the boats in the fleet to see if he might measure up for a job as a deckhand.

"Would I be able to go out with you, Cap? I promise I won't be in the way or be any trouble."

I said, "I'm not sure that you'd like it. The weather doesn't look good for the next couple of days, and the seas can get quite rough. You need to know that I never turn back once I start a trip, no matter how sick a person gets."

"Oh, you don't have to worry about me, Cap. I'm a 'Tin Can Man.' I've sailed on destroyers out of Newport for three years, and I've never been sick," he boasted.

Well, he seemed determined, so I gave him the time that we would be shoving off. "I'll not wait a minute if you're late," I warned him and began to walk up the dock toward my pickup.

The next morning, right on time and well before dawn, Old Swabby showed up. He threw his sea bag on deck, and I had one of the crew show him to the bunkroom in the fo'c'sle. "Listen, Moe," I said, "put him on the top bunk way up forward, and then close the door."

That done, we headed out and, true to my prediction, we began burying the bow as soon as we got clear of the whistle buoy off the Point. Conditions stayed bad for nearly seven hours. When we made our bearings and set out the net, it was coming on daylight.

Because of the rough water, I made the first tow a short one so that we could get an idea of just what might be on the bottom before committing ourselves. When we hauled back and the doors broke water, the net's cod end surfaced like a whale coming up for air. We then began "splitting"; that is, tying off one segment of the net—about five thousand pounds' worth of fish—and emptying that before attempting the next one. After the second hoist came aboard, there was a foul-up in the splitting strap.

Red hake and whiting have air sacs in their bodies, so our very full net was floating high enough to allow me to run out on top of the fish to straighten out the splitting strap. With the sea making up in good shape, it was a pretty tricky maneuver but it had to be done. Naturally, my boots would start to sink into the fish, so I had to keep moving quickly. There was also the danger that the floating net might roll me off and into the churning sea, slamming me into the hull.

As soon as we got the net back overboard for another set, and all the fish were put below in the hold, I sent Moe below to see where our "old navy" man was. I had not seen him since he stepped aboard in Galilee.

Moe came up to the wheelhouse looking kind of pale and said that the fellow was still lying in his bunk, periodically gushing vomit all over himself and not moving as he did so. I figured that if our guest was still upchucking then he was still alive, so I had Moe again close the door to the forward bunkroom.

We completed three tows and filled the hold with seventy-five thousand pounds of fish, and we had one huge cod end with about seven thousand pounds in it sitting on deck. We then hauled the drag doors aboard and secured everything for a long, wet trip home.

The navy boy was not seen or heard from for the entire trip. When we got into Point Judith and tied up at the processing plant, euphemistically called the "trash plant," I sent a couple of the boys below to rouse the old salt. After a long pause, all three emerged from the companionway, my men practically carrying our poor guest, who reeked of vomit worse than anyone I had seen before. He never said a word but climbed onto the dock and staggered toward the parking lot where he had left his car. We never saw this fellow again.

I then sent my men down to drag the mattress and bedding out on deck and put them into a garbage dumpster. Then they took the deck hose below and flooded the bunkroom, pumping all the residue overboard as fish bait. Without a doubt, that was the very worst case of "saltwater blues" I had ever encountered. But it certainly wasn't the last bad one.

One beautiful summer day when my son was about six or seven years old, we were steaming to the east'ard on the *Dorothy & Betty II* with a good friend, who happened to be my insurance agent, on board. Dick Soderberg had insured many fishermen over the years, but had never experienced the work they engaged in. When he asked if he could go out with us, I was delighted to agree—with the same stipulation that I gave everyone else: "I won't break up a trip in the event of seasickness." Dick had accepted that and we had scheduled him for the next trip.

Prior to our leaving Point Judith, we had received a weather report about a hurricane that was working its way up the coast. The report said that on the day we were to leave we would begin to feel the effects of the heavy ground swell preceding the storm.

The report was accurate, as seas were breaking over the seawall, although the surface of the water was as smooth as glass and would remain so.

The first part of the trip was uneventful. Dick seemed to handle the rollers quite well, and both he and I were pleased—he because he felt fine, and I because I didn't want to feel guilty for being hard-hearted in the event of his becoming sick. Things continued well until we were heading home. The seas were still glassy calm on the surface, but the swells were increasing in size and frequency. The sun was just glowing in its beauty, and the sky was clear, except for a few wispy clouds, as you would expect.

The cook called out that dinner was on the table, and all hands were anxious to go below because we had not eaten since a very early breakfast. The time was now 2:00 P.M., we had handled a boatload of fish, and we were ravenous. I sent everyone below to eat while I held the wheel. Dick, however, declined my invitation, as he was just starting to get a little woozy and wanted all the fresh air that he could get. I had just commended him on his ability to handle the heavy seas that we had encountered since leaving the Point when my son David came up the companionway. Eating an ear of corn, he climbed up onto the fish-hold hatch covers. He was rolling and swaying in concert with the motion of the boat in the heavy seas—forward, then sideways, then backwards, again and again, all the while eating his ear of corn. Butter and juices were running down from his mouth and dripping onto the deck.

All this time, Dick couldn't take his eyes off David, and his pallor began to change from snow-white to pale-green; then he lunged toward the rail of the boat and exploded his innards into Davy Jones's Locker. When he had expelled everything that he had eaten for the previous ten days, he collapsed onto the deck.

While this was going on, Davy kept munching on the ear of corn, totally unaware of the distress it had caused. Dick finally gathered up enough strength to climb back onto the engine-room trunk, where he had been perched for several hours. He gazed wistfully at David, looked up at me, and said weakly, "Your son is a great little sailor, isn't he?"

ALONE WITH THE SHARKS

It was the first week of July in 1951, as I recall, and I was about to begin my first year aboard the *North Wind*. Jerry Adams owned her and was captain.

I had been slowly working up in the ranks, if you will, from day boats—those that left for local fishing grounds early in the morning and returned late in the afternoon, just in time to unload their catch at the Point Judith Fishermen's Cooperative. The crew would then repair gear and get the boat ready for another early start the next morning. Some of these boats, like the *Jane Lorraine*, the *Joyce Ann*, and a few others did really well. But I was the new boy on the docks at Galilee, and all of my experience prior to leaving school was as a trap fisherman. Since I hadn't been dragging for bottomfish, I had to begin with any skipper who would give me a chance.

The farthest boat trip I had made when I was young involved sailing around Point Judith to the fish traps located off Narragansett and Black Point. I could mend nets, splice wire, and handle myself on the deck of any vessel, so, while I wasn't a novice, I remained an unknown quantity to the men of Galilee.

To prove myself an able and hardworking hand, I began the long haul to success by going to work for the first captain to offer me a site—Marty. He was a unique individual. He came from a family of some means and probably could have settled down in a job ashore and been very successful. He was intelligent enough; he just lacked a few qualities like drive, ambition, and any interest in success. You see, Momma was always there to bail him out of a jam. The most frustrating part of this was that Marty was an outstanding fisherman. Every day that I could persuade him to leave Durfey's coffee shop, get away from the other ne'er-do-wells swilling java and swapping lies, and go fishing, we did really well.

Another problem was that Marty was single, so a good day's fishing set him up for a week or two. By contrast, I was a newlywed and needed a steady income. So with a few months' experience on Marty's dragger, I moved on to bigger and (I hoped) better things.

Gradually I gained enough experience and had been observed long enough by different skippers that I progressed to better, more productive, and more seaworthy boats. Finally an opening became available on the *North Wind*, which was known as a highliner. One of the men was getting married and wanted to take a trip off. I heard about it and went to see Jerry, saying that I would work without a share (wage) if he would just give me a chance to prove myself.

I think my approach caught him off guard, and he said that I could sign on—but only for the one trip. Such temporary substitutions were a common event. Many crewmen were transients, some by choice; that is, they would be available to ship out at a moment's notice if a permanent crewman was sick or had some reason to stay ashore for a few days. I did not plan on being a transient for long, but I would have agreed to almost anything to get aboard a highliner like the *North Wind*.

We headed south from Point Judith on that first day as the sun began to rise over the eastern horizon. The wind was calm, and there was a gentle swell from the sou'west—a perfect day for spotting swordfish. We would be running the fifty-fathom curve from east to west, as this was the most likely depth at which to find swordfish early in the season.

All swordfishing carried on at this time involved harpooning. Long-lining—fishing with hundreds of baited hooks on a line that stretched over miles and was buoyed and anchored at both ends—was unknown on the East Coast, as was the use of an airplane to swordfish.

When the fish first showed themselves in our neck of the woods, it was in an area just south of Block Island. They were small, weighing between one hundred and one-hundred-fifty pounds dressed. They were bright blue on the back, quickly becoming silver and then white toward the belly. The young fish were always moving much faster than the older ones as they swam, and they were more difficult to harpoon as a result.

As the season progressed, the first and youngest fish would move to the east'ard, and the next bunch would range between one-hundred-fifty and three hundred pounds. Moreover, they would be a darker blue on the back flowing downward to lighter blue rather than silver and white. They would also be slower in the water and would hold their position for a longer period of time, but we would not usually find them until we fished south of Martha's Vineyard and east to Nantucket Shoals.

Once the fish neared the shoals, we would lose them until they showed up on Georges Bank, which ran east of Nantucket and north toward Nova Scotia. These would be the big, chocolate-brown fish ranging from four hundred to eight hundred pounds.

Jerry Adams was perhaps the best striker in Point Judith and, other than Bijarny Larson from Martha's Vineyard, perhaps the best in the Northeast. So I was confident that if the young fish showed themselves, we stood a good chance of getting our share. That first day was a broker—no fish at all—but, this early in the season, that was to be expected. That night, we sailed into Old Harbor, on Block Island, as the wind was freshening from the sou'west, and we were not sure what the next day would bring.

As things turned out, the weather was as good as on the first, and as soon as the sun was up high enough to warm the surface, the fish began to fin. Jerry hit the first one, and Art Raposa, the *North Wind*'s number-one doryman, tended it. Before Art was able to bring it to the surface, we struck another, and that was my fish to tend. I quickly slid down the backstay of the mast with my hat around the cable to prevent a burn on my hands, landed near the stern of the boat, pulled the lanyard leading to the dory, and jumped in before the dory had time to drift away from the mother ship. As soon as I landed, I threw my weight to the starboard side, and this brought the dory to the right, allowing me to intercept the keg before it had a chance to be towed very far by the ironed swordfish.

Quickly catching the keg with my gaff hook, I brought it aboard, then settled down on the forward thwart, inserted a fairlead roller into the tholepin hole, laid the warp into the roller, put a bushel basket between my knees, and began to haul the warp stretching between the keg and the fish, coiling it into the basket as I hauled. I kept up steady pressure, careful not to yank hard, and as the fish would run, I grudgingly allowed the warp to slide overboard. As soon as the fish tired a bit, I would haul as fast and as steadily as I could.

Within twenty minutes or so, I had the swordfish alongside the dory. By holding a steady strain on the warp, trapping it against the rail with my knee, I would still be able to let it go in the event the fish began to dive. I then took a heavy nylon strap with spliced loops at both ends and worked it around the body of the fish just ahead of the tail. Cinching up the strap until I could slide one loop inside the other, I pulled the tail out of the water.

Once this was done, I wrapped the strap around the rising (the wooden strip of wood running the length of the dory along the gunwale) while holding the warp, meaning that the swordfish was unable to thrash enough to get away. Laying my weight back against the side of the dory opposite the ironed fish, I raised it higher out of the water and lifted my oar. That was a signal to the crew on the *North Wind* that I either had the fish or was in trouble. In either case, they would respond as quickly as possible.

As the *North Wind* drew near, I picked up the "gilling iron," a wooden shaft about four feet long with a flat iron probe about two feet long attached to it. It looked much like a butter knife in profile and was not sharp on the edges. My job was to run this tool into the gills on either side of the swordfish's head, probing as I went, causing the fish to bleed out.

A swordfish has a great quantity of blood, and as mine lay in the water and worked his gills in an attempt to breathe, he pumped that blood out into the ocean around him. This allowed all of it to be purged from the system and would make for a better-quality fish once it was aboard the mother ship and iced down.

When the *North Wind* came alongside my dory, with the fish between us, I reached for the "whip." This line, which had a strong cast-iron hook on the end, ran upward to a block overhead and then down to the *North Wind*'s winch. I inserted the whip hook into the strap secured around the tail of the fish. As Jerry hauled it aboard, I passed the keg and warp to one of the crew, then slid the dory aft, where I could again tie it off until the next fish had to be tended.

We struck three swordfish that day, bringing them all safely on board, and as we sailed back to Point Judith, Jerry said that I had done a good job. This was high praise for a novice deckhand to hear on his first day, which—in my case—would be my last aboard the *North Wind*.

But while we were unloading the swordfish and picking up ice and fuel for the next trip, Jerry heard from the crewman that I had temporarily replaced. It seems that being a newlywed, he wanted still more time ashore, so he had decided to work with his father on a small day boat. This would bring him ashore each night, albeit a short night, but a night at home nonetheless. Jerry then asked me if I would like to stay aboard the *North Wind* for the summer, followeing the swordfish as they progressed to the east. Of course, I jumped at the chance.

We had a successful and an uneventful season, fishing from Block Island and moving ever so slowly eastward each time we ventured from the Point. Gradually we lost the main body of swordfish on Georges Bank, off the east coast of Massachusetts.

It was time to head up to Nova Scotia. This being the case, we first had to go back to Point Judith, unload the fish we had on board, and gear up for a much longer trip. We cleaned the hold extremely well, even scrubbing it out with Clorox to kill any lingering bacteria. We then laid three-hundred-pound blocks of ice on the bottom of the fish pens and floor of the hold. These would last much longer than the crushed ice we used for shorter trips. On top of the layers of block ice, we put tons of crushed ice to pack around the swordfish as they were stowed belowdecks.

The cook would store all of his meat and dairy products in one of the fish pens, packing the food in the order in which he would use it. The more perishable items, like chicken, would be topmost. When we got down to the beef roasts, late in the trip, we often found them green with mold. The cook would thoroughly wash the meat with baking soda and put it in the oven anyway. By using these precautions we never, in all the years I was at sea, had problems with the food.

I can't say that about the drinking water, though. It's a wonder any of us survived what we drank. We had water tanks that held about three hundred gallons, and with six men, that water was quickly used up. We also had a few fifty-five-gallon drums—fuel drums that had been flushed out as well as possible. Still, the water they held had the taste of fuel when we drank from them. We tried to reserve that water for what little bathing we could do, and for cooking and making coffee. I think they call that particular flavor "café latte" today.

All our swordfish kegs, with the exception of a half-dozen to be used on quick notice, were filled with fresh water and bunged. These were new kegs made from pine, and in no time the water took on a greenish cast and a malodorous smell. These were the very last containers to be opened, and then the water was boiled before use. Even then it was unpleasant to drink.

Great quantities of canned food and dry goods were stuffed away in lockers in, under, and around every structure in the fo'c'sle, wheelhouse, and engine room. In the remaining areas we stowed our gear, personal items of clothing, and the like.

The most unpleasant condition was the lack of water to bathe in and in which to wash our clothes. Let it be said that you would never want to cross us to leeward after we had been to sea for a few weeks. Fishing boats today have flush toilets (instead of the five-gallon buckets we used) and hot showers, and they offer a nearly inexhaustible supply of fresh water.

On our first Nova Scotia trip, all of these preparations took only a day and a night. We had to make haste as we were coming into the hurricane season and would be a very long way from home. The next day we had all of our families come down to the docks to see us off. It would be several weeks before we were back home. With all of the expenses incurred for this trip, we did not plan to return until we had a hold full of swordfish.

Having said our good-byes, we sailed out of Point Judith and headed east toward Nantucket Sound and Monomoy Point. From there we headed northeast toward the Peak on Brown's Bank and Wilkinson Basin, just south of Nova Scotia.

With a southerly breeze and a following sea, we made excellent time, but it was still the third morning after leaving home before we went aloft to begin our search for the ever-elusive and fast-moving swordfish. We began a tacking (zigzag) pattern to maximize our coverage of the ground we were fishing as we moved ever closer to the peak. The first two days, we saw nothing at all.

In the forenoon of the third day we came across a chocolate-brown swordfish that looked to weigh about four hundred pounds or so. Jerry struck him, and he disappeared. The skipper called to me as I prepared to get in the dory and said that the fish was hit hard, but due to his size, I would need some time to haul him in.

I jumped into the dory, pulled the painter free, and began easing toward the keg, which had not yet begun to move. Apparently, the fish had not yet sounded, which gave me some time to get settled.

Once I had the keg aboard the dory and put some tension on the warp so that we would know which direction the swordfish was heading, Jerry came alongside and told me that he was going to start steaming in a spoked-wheel pattern, using me as the hub. By doing that, he was more likely to pick up the main body of fish. Jerry's last words were "It'll be a while before we get back to you, so be careful."

The sea was flat calm, there was no wind or fog, and the fish was well ironed. What could go wrong?

I began to pull steadily, taking in the slack when he gave some and holding a firm strain even when he tried to pull the warp out of my hands as he dove or ran in a different direction. I kept wetting my leather gloves to prevent my hands from getting burned as the warp ran out time after time.

Slowly I made progress in hauling the fish closer and closer to the surface. The water was crystal clear as I looked over the side, and I could see the fish struggling against the ever-upward tension. Obviously, it was beginning to tire. I knew that I was, but I could not rest. If I stopped, even for a moment, the fish would regain its strength and dive again. I had to keep up the constant pulling against the downward thrust. My back and leg muscles were by now writhing in pain. My arms and hands were nearly numb. Finally I took one more look over the side to see how close I was to having him at the surface. Just a few more feet. Then I saw something that chilled me to the bone: sharks! I did not get a chance to see how many, but even one was too many.

It was common to see sharks around a wounded fish, especially if it was bleeding, but you never knew how they were going to act. Blue sharks would usually circle around slowly—unless they were in a feeding frenz; then they would move straight in toward their prey. As long as you could see their fins on the surface, you could feel some comfort that they were not about to attack. They were narrow in the head and had sharp, pointed noses.

The sharks that terrified me were called "gray porkers." They had a wide head and a shovel-shaped nose. You would very seldom see any signs that a porker was around until it struck. I've seen them come up from the depths as we were hauling in a cod end full of butterfish, and before we could get the bag out of the water, a porker would appear to bump the net, then roll away in an effortless turn. Out would stream the butterfish through a hole the size of a bushel basket.

The cod end was made of tough nylon parachute cord, but it would give way to the shark's teeth as if it were a paper bag—a wet paper bag. Scary? You bet.

An account related by some fishermen involves a porker that attacked a swordfish that was lying in the water after being harpooned head-on and struck dead. The shark swallowed the head and sword, and kept on swimming. Yes sir, they made me nervous.

I brought the swordfish up alongside the dory and strapped him down before he could get his second wind. Because the *North Wind* was nowhere in sight and sharks were swimming all around me, I decided not to use the gilling iron. I would wait until I was picked up by the mother ship. The day was a beautiful one, so I leaned back against the opposite side of the dory from the tethered fish to wait.

About fifteen minutes later, I saw the *North Wind* steaming right toward me. I raised my oar to signal that I had the fish. The boat came straight at me, and when I could see everyone in the rigging looking at me, or so I thought, I ran the gilling iron into the swordfish's gills. Blood began to surround my dory, and the sharks were coming ever closer. *Come on, Jerry, hurry up,* I thought. Just then the *North Wind* made a tight turn to port. One of the men aloft was pointing off toward the horizon as if he had spotted another fish. In a matter of minutes the boat was merely a speck, at least from my position in the dory.

If I was to save my swordfish, I realized I was going to have to roll him into the dory with me. I had done it before, but not in this dory and not with this size fish. I quickly ran some ideas around in my head. If a dory is built right, the theory is that even if you swamp it (fill it up with water), it should stay right side up for an indefinite period of time. And, if you should happen to roll a dory over, you should be able to stretch out on its bottom, again for an indefinite period of time. *Well, Cottle,* I thought, *let's test out those theories.*

The fish lay dead in the water, so I removed the strap from the rising, laid the dory on her side until she began shipping water, grabbed the fish by the gills and tail, and rolled it into the boat as I threw my weight to the opposite side.

My aim was to gain enough leverage to "flip" the swordfish aboard. Then, even though the dory would be low in the water, I should be able to bail her out once I stabilized the situation.

The fish started in as I had planned, but then his dorsal fin got caught in the rising, and I could neither get him in nor out. Over went the boat, dumping me in the water. Quickly, I pushed the bleeding swordfish as far away from me as I could under the circumstances. I had gone as far as I was willing to go to protect that fish from the sharks. Now my first priority was to put wood between me and those eating machines from the deep.

I rolled the dory upright, although it was full of water, and scrambled into it as quickly as I could move with wet clothes and boots. I sat amidships, put my hands on the gunwales, and was content to wait until being rescued. That seemed like a fine plan, but in this instance, it didn't work. Just as I settled down on the dory's bottom, it rolled over and threw me out again

As I went overboard, water pressure sealed my boots tight to my legs, and I started to tip upside down due to the air in them. But, can you imagine? I was angry not with the poorly built dory that was threatening my life, but with the idea that I was going to be forced to get out of my boots—my brand-new, eighteen-dollar boots! Once I had accepted that decision, however, I now was forced to figure out how to get those vacuum-sealed things off my feet and legs. During one brief period inside the dory I was able to remove them just as the boat rolled over again. As I fell out, I could see the boots sinking slowly beneath me. They probably became a meal for the old porkers; they'll eat anything.

As I repeated the effort of climbing into and being thrown out of the swamped dory, I tried to create as little disturbance as possible. Nothing will draw the attention of sharks any more quickly than the sound and thrashing of a wounded fish. And besides, I had to stay within reach of the dory, as I did not know how to swim. Not then, not now. I always figured that where we fished, the nearest land was six hundred feet away, but straight down. Swimming wouldn't get me there.

Soon tiring of the never-ending struggle involved in rolling the dory over and being cast out, I reconsidered the theory: A well-built dory, if tipped upside down, would indefinitely hold a man who stretched out on her bottom. So that had to be my next maneuver, and it was easy enough because the boat did that by itself. Getting up onto the dory's bottom, however, was another matter.

I finally accomplished my goal and was lying there trying to catch my breath and get some rest. Suddenly, the dory lowered her bow into the water and slid me off as quick as a wink. I clambered back up only to be dumped again. Over I went, my fearful mind on the sharks while I was struggling with that miserable dory.

How many times I went through this process and how much time went by, I don't know. All I know is that I kept trying, and finally lost all awareness of the futility of my situation. I concentrated on my struggle with the dory. I forgot about the swordfish; I forgot about the *North Wind.* All I thought about was, "When are those porkers going to hit? Where will they hit? My legs? My gut?" I just *had* to keep the wood of that dory between us.

Ultimately the *North Wind* headed my way. Jerry could see the capsized dory and the keg and oars floating nearby, but he could not see me. "My God!" he cried out. "What will I tell Gloria?"

Just then one of the men aloft called down from the mast, saying that I was on the far side of the dory, but all they could see was my right arm waving in the air. I have no recollection of the men picking me up out of the water. I do remember watching as they hauled that damned dory on deck and refitted her with gear, Jerry asked me if I wanted to have one of the other fellers haul my fish in, but I told him that I would do it.

"Let me get on a dry set of clothes first," I said.

Meanwhile, the crew relaunched the dory, towed it up to the keg, and scooped up the oars floating nearby. When I came back on deck, I jumped in, grabbed the keg, and quickly hauled the dead swordfish to the surface.

As I put another strap around his tail, Jerry passed me the whip, then hauled up the fish and set it on the deck.

After all of the crazy business involved in trying to protect that swordfish from the gray porkers, there was not a mark on him other than where he had been ironed. I had learned two lessons: First, to hell with protecting swordfish from sharks; and, second, so much for theories on dory seaworthiness.

DANGERS FROM AFAR

It was the spring of 1958. We were fishing along the continental shelf—or, as we called it, the Gully—in one hundred fathoms of water. The day broke bright, clear, and warm for so early in the season. But then again, we were enjoying the warmth of the Gulf Stream coming up from Florida, and it would continue to meander northward to Nova Scotia, past Labrador, and on around Ireland and England.

During the night we had enjoyed a decent catch of butterfish, but with daylight coming on we knew they would rise up off the bottom. So, before we had set out our most recent tow, we had changed nets. The mouth of the butterfish trawl was chained lightly along its bottom edge—just enough to tickle the seabed—because butterfish normally swim close to the bottom but not hard against it. We also had upward of twenty-five or thirty "cans"—round, heavy-duty, aluminum floats—lashed to the headrope at the top edge of the trawl mouth. (If we did not use rugged, welded cans at that depth, they would be crushed like a beer can in a teenager's hand as he tries to impress some young lady.) This allowed the net to rise much higher in the water as we towed along, hopefully through the schools of butterfish.

When "on bottom," looking for tilefish, fluke, or lobster, we used a net whose footrope was weighted down with larger links of chain—and more of them—allowing it to settle *into* the seabed. We also used fewer cans, as we did not need to raise the headrope off bottom quite so much.

At any rate, we were working on our last series of tows. We had been out for three days and nights, and these final hours should top the trip off nicely. If we fished until dark and then headed for home, we could be in Point Judith, tied up to the Fishermen's Co-op, in time to catch the morning activity at New York City's Fulton Fish Market, which pretty much controlled the daily price of fish along the entire northeastern seaboard. Besides, I didn't like to see too many sunsets at sea before enjoying the next one from my back porch at home.

We had been into the morning tow about half an hour, when I picked up a large blip on my radar screen. Some ship was coming on fast—much faster than any fishing boat I had seen on these grounds. By the time we had hauled back, dumped the catch on deck, and reset the net, this "stranger" was slowing down and had come within a quarter-mile of us. Man, that vessel was like no fishing boat that I had ever seen before! She was at least two hundred feet long, and a lot of time had passed since she had been hauled out and had experienced the thrill of a wire brush and a dab of paint.

I have never seen as many antennae sticking out of a wheelhouse or lashed to the after mast as were attached to this old "slab." On the huge funnel rising out of the engine room aft were the hammer and sickle, and the red star of Mother Russia.

No one aboard the *Dorothy & Betty II* had a camera, so I jotted down a description of the vessel and whatever numbers I could make out. When we were about to complete our second tow, the Russians began to put their drag doors overboard in preparation for setting their net. Well, I had never seen such a "rigging" (awful mess) in my life. The Russians had more power and equipment on board than all the fleet at the Point, but I'll be damned if they knew how to get those doors off the deck and over the rail. My crew had hauled back, brought our fish on board, cleaned the deck, iced the catch below, and made ready to set again while the "Russkies" were still messing with their doors.

As Red Skelton would have said, "It just don't look right to me." I told my crew to haul the wings of our net aboard; I wanted to watch this circus for a while. Any professional or tradesman who observes a man claiming to be of his own calling takes no more than a few minutes before being able to tell if someone is legitimate or not. The boys we were watching were *not* fishermen. One man on deck, probably the mate, was yelling and scurrying forward and aft, trying—without success—to get the crew to follow his instructions. I really felt sorry for him. The crewmen must have all just come off the farm.

I put our boat in gear and made a slow turn in the direction of the Russian vessel, as though we were getting into position to set out our gear.

When we came within hailing distance of them, I stepped out on deck and called out a greeting of some sort—I can't remember just what—and waved at the same time. The skipper of the Russian boat stepped out on the bridge of their wheelhouse, and he in turn waved and shouted a greeting. While I was alongside, I put in a call to the Coast Guard at Point Judith. The radioman responded immediately, and just as I pressed the button on my mike to give him a description of this foreign ship, the Russians pressed some device on their radio and blanked me right out. I kept trying to get my call through while putting more and more distance between us, but the power of their set overwhelmed us. I finally stopped steaming and told the men on deck to haul all of our gear aboard, including the doors. We secured everything for the long steam home, and because we might yet encounter some heavy weather, we battened every thing down.

We were aware, as I'm sure the Russians were, that we were in the traffic lane used by nuclear submarines as they shipped out of their base in New London, Connecticut. Once the subs found deep water, they submerged and went in any direction as they patrolled the seas of the world. This charade that the Russians were putting on, pretending to be innocent fishermen, put them in a perfect position to track the American subs. My effort to warn the navy was not what they wanted—hence their jamming of my radio transmissions.

We steamed homeward for over four hours before I was able to contact the Coast Guard again. When I finally raised them, I described the incident and gave them as much detail as I could.

Since it was nearly midnight when we arrived at Galilee, we tied up and went home. The next morning, while we were at the Fishermen's Coop unloading our trip, a couple of naval officers arrived and asked permission to come aboard. I went back into the wheelhouse with them and reiterated what I had passed on to the Coast Guard. One bright, young, but obviously innocent officer said to me, "Why didn't you ask to go aboard? Then you would have been able to look around."

My response was, "For one thing, the only Russian word I know is 'vodka,' and for another, I had no desire to be called a spy and end up in some Siberian prison." The officer made no further comment. After I completed my report to the older, and apparently wiser, officer, they thanked me for the information. They said that the navy would be sending a PV2 observation plane to watch over our new neighbors.

This was my first encounter with a foreign ship of any kind, at least on our fishing grounds, and I had a feeling in my bones that this would not be the last such incident. I am sad to say it was not; and in one case it ended tragically.

Within a month or so we had half a dozen large Russian trawlers fishing the area that had always been worked exclusively by the American fleet—and by not so many of us at that. Now, powered by massive engines, these big, heavy foreign rigs were raking the bottom with nets many times the size of ours.

While the American fleet had to go to its home ports every few days to unload and refit, the foreign trawlers stayed on the grounds week in and week out. When their holds were full, they called their mother ships to come alongside to unload their catches and refit them with food, fuel, equipment, nets, ice, and women. Yes, we've run alongside some of these mother ships, and women were waving and calling out as we steamed by. To tell you the truth, they were as big and strapping and homely as the crew they were servicing. But all in all, the foreign fishermen were tough competitors. They had every advantage.

Soon the continental shelf off New England looked more like the parking lot at the United Nations building, complete with ships of every flag: East Germany, Russia, Iceland, Norway, and others. They would set out their nets and then run their tows end for end until they sailed over the horizon and out of sight, only to reappear again in the next day or so. They were collectively catching everything that swam along the bottom. Soon, they carried their ravaging of the sea even further. New ships called "pair trawlers" that used one huge net, a midwater trawl towed between them, showed up on our grounds. As the name implies, this net was lowered to a particular depth where schools of fish had been seen on echo sounders that were much more advanced than ours. Then, with the towing power of both large vessels, they would work that net toward the school and quickly gather it in, unloading the catch into one of the two boats. Were they effective? Consider this:

In the early sixties I had been working with a governmental agency out of Woods Hole, Massachusetts. As I steamed offshore, and again on the homeward run, I would periodically stop and lower a torpedolike instrument to record the ocean's temperature as the device was lowered and again as we raised it back aboard. Doing this work brought me into contact with some of the agency's field men. One of them, Pete Couture, would tell me of their findings, whether they involved water temperature or fish stocks.

One day he told me that there was a school of "round herring" that started around Nantucket Lightship, off the Massachusetts coast, and ran all the way down to Hudson Canyon, off New York City. Within a few months of the arrival of the midwater trawlers, however, that school—of, what, millions of tons, perhaps tens of millions of tons of herring?—was gone, scooped up, never to be found again.

Not long afterward, as a director of the Point Judith Fishermen's Cooperative in Galilee, I had the opportunity to meet with the assistant secretary of the interior in Washington, D.C., and I posed a question about the impact of the foreign fishing vessels then working our waters. The secretary's response was, "One missile base in Iceland, or any other NATO country, is worth more than the whole American fisheries—on both coasts." I respected his honesty, if not his answer. So it was up to us to weather the situation as best we could.

The adverse impact on our fisheries was serious in that it affected our overall economy and our ability to provide for our families. But the problem extended beyond that. A particular aggressive attitude within the Russian fleet was becoming more and more evident as time elapsed. At first it was merely a display of poor seamanship. Just as on highways ashore, there are rules at sea. When ships appear to be approaching head to head while steaming, each should quickly yield to the other by turning to starboard, thus widening the distance between them as they pass. This provides greater safety and is a display of common courtesy between ships.

A similar courtesy is extended to a vessel that is towing a net. The standard within the American fleet of eastern-rigged draggers in the 1960s was that the net, doors, and cable were hauled and set off the starboard side; this way, the boat could turn quickly only to starboard, in the direction of the gear.

This was especially true when fishing the continental shelf, where the tides are extremely strong. If a boat got the tide on the port side, it might turn the vessel right around, causing a twist in the tow wires and eventually tangling the doors and net. That would be a real nightmare in the middle of a stormy night with worsening sea conditions. It certainly could create a situation where men might be injured or killed. This situation would impact all smaller fishing boats in the range of fifty to eighty feet long with relatively limited engine horsepower. Knowing this, everyone was especially careful in how they approached an eastern rig that was towing. This could be quickly discerned in daylight by observing the vessel laying to, turning, or making some other distinctive maneuver. Then, too, a vessel that was towing would have a metal fish basket aloft in the rigging. At night, we had a system of lights on our masts that quickly indicated that we were towing, allowing all to give a wide berth.

The Russian trawlers, as well as all the other foreign vessels, were gifted with engines developing thousands of horsepower; therefore, they were not limited in maneuvering as we were. Yet they gradually began to crowd us when fishing in the same area. Some were more aggressive than others. A number of times I had to turn out of my right-of-way and even begin to haul the net back, because if I continued as headed I would become entangled in the other boat's trawl gear and get twisted into a nightmare of nets, doors, and cable. Being of reasonably sound mind, I would yield my position in an effort to prevent a critical problem from developing.

One time a Russian began to crowd my brother, John, aboard the *Four J's,* a beautiful schooner hull of sixty-five feet, as he was towing. John got on the radio channel that all fishing boats were tuned to and asked the Russian to turn aside, as the American boat had the right-of-way. His request was ignored. He then transmitted his message again, even as the Russian was getting ever closer. John warned the Russian captain that he was putting himself in harm's way by his actions. There was no response. When the Russian was nearly on top of the *Four J's,* John loaded up his pistol with one shell in the chamber and ten in the clip. He strode out on deck, stepped up onto the hatch combing, held the gun up in clear sight for a moment, then fired all eleven rounds into the wheelhouse windows of the Russian ship.

Can you just imagine the chaos that those eleven bullets caused as they ricocheted around and around inside that steel wheelhouse? Well, let it be known that the Russian was quick to yield, and John continued to tow in his chosen direction. He then keyed the radio mike and, in a very magnanimous voice, thanked the Russian captain for his kind gesture in yielding.

It became obvious to us all that a more serious event would take place sooner or later. We just didn't know when it would happen or whom it would involve. That day, and that event, came much sooner than anyone imagined.

On a cold February night in 1959, there were probably a half-dozen American boats fishing "the edge," some from Point Judith and the others from New Bedford, Massachusetts. Of that small fleet, the *R. W. Griffin* out of New Bedford was by far the biggest, at about eighty-five feet long. She was a handsome eastern-rigged dragger—deep in the belly, "sea smart," and according to her captain and crew, as comfortable as any boat of that size fishing on the shelf.

The night the much-anticipated incident happened—nearly a year after the incursion of the first Soviet spy ship and, shortly thereafter, the USSR's fishing fleet—was wicked cold, with a wet easterly wind. Though the breeze hadn't yet kicked up, it had all the makings of a bad blow. The glass was dropping slowly, and the wind had backed from southeast to nearly northeast.

This was the kind of weather that was hard to fish in, but it usually drove the butterfish down to the bottom, where we could reach them with our nets. I had called my brother to see if he had seen signs of fish, but he hadn't come across any. Both of us knew that this would be our best opportunity of the trip to strike a large body of those elusive silver critters.

It seemed that our best catches were preceded by an easterly breeze accompanied by rain and increasing seas. The more miserable the weather, the better the fishing. Trouble is, a large haul of butterfish is very heavy. We have caught as many as twenty thousand pounds in a short (half-hour) tow. When you have that much weight in the cod end and the seas begin to increase, there is a much greater chance that the net will tear—sometimes in half—releasing everything inside the twine as well as creating a lot of mending work.

We tried a few tows in different depths without success, and with the weather getting nasty, I decided to change nets. We unshackled the butterfish gear, dragged it up forward, and secured it ahead of the doghouse, where it would be out of the way for the rest of the trip. Now the seas were breaking across the rail every so often, which made working on deck a real effort.

When the net we used for true bottomfish was rigged up, we eased it overboard and began a turn to get us moving in the general direction that I wanted to tow. I blew the air horn, and the crew eased the doors into the rolling sea and they headed toward the sea floor, some one hundred fathoms below us.

Two hours later, a heavy chop was throwing spray across the deck as we turned into the wind to haul back. Just as our doors began to break water, I looked up to see the *R. W. Griffin* towing by. He was just a bit inshore of us, but I knew the bottom would have risen to about eighty fathoms there, as the mountain edge on which we were fishing was quite steep. Two boats could be within throwing distance of each other, yet there could be a difference of ten or more fathoms in the depth of the water beneath them.

The other vessel's captain and owner, Warren Griffin—Sam, as many of us called him—was a rugged man whose gray hair was beginning to turn white and whose voice when he spoke over the radio was strong yet pleasant. He had just called, and since I was too busy to answer, I flashed my deck lights a couple of times, and he acknowledged over the radio that he understood my message. We would have the whole evening to chat as we both ran a "six-and-six watch"—that is, the captain held the watch from six at night until midnight and then turned in until six the next morning. Our first mates ran a similar watch, while the crew usually stood an "eight-and-four watch." Since our tows were usually longer at this time of the year, when we were bottom-fishing, I had ample time to talk with other skippers once the deck was clear and the crew was belowdecks.

That night, the fleet was strung out in a line five or six miles long, making one tow after the other. For a while, we would see each other's running lights or deck lights, then they would disappear from sight—especially as the weather worsened. All of us had radar, though, and we were able to keep track of one another, as well as to watch for other traffic.

Once we had set out the net, and while my crew was picking up the fish and lobsters and icing them down in the hold, I gave Warren a call, but it was his turn to be hauling back. I turned my attention to the radar and was trying to locate just how far away the *Griffin* was when I saw a much larger blip on the screen. Because of its size, the vessel was without doubt a foreign trawler. Because these vessels had so much power, they could tow a net as fast as most of our boats could steam, so it was difficult to tell if this one was fishing or not. At any rate, it was closing fast on the *Griffin*, and since Warren was hauling back, he was probably not aware of the approaching trawler.

Even if he did know, there was little he could do. Once a dragger's doors and net were clear of the bottom, the crew had to keep everything coming up or they would end up with a terrible tangle of gear, as well as facing the danger of putting the wires and/or net into the propeller. This far offshore and with bad weather edging in, that would create a very dicey situation. It was just because this kind of problem could befall any one of us that we were very careful to look out for each other and were quick to make room for the other fellow if we possibly could.

The closer we came to the *Griffin*, the more dangerous the situation appeared—even though I was watching the evolving incident on radar. From the movement of one blip, it was obvious that the *Griffin* had hauled back and was laying off the wind as the doors and then the net came up. Warren was undoubtedly out on deck, ready to hook up the after door once it reached the gallus frame. I knew that he was unable to see what was about to overtake him.

The second and much larger target on the screen was steaming downwind toward the *Griffin* and would strike her on the starboard side, just forward of amidships. Suddenly, the radio came alive. Warren must have taken a quick look at the radar screen and seen the blip that was coming at him out of the wet, easterly sea. He had just enough time to call out "Mayday! Mayday! Collision, collision! Russian . . . "

That's all I heard. Shortly after this transmission, a huge foreign trawler towed by us without any comment, concern, or interest in what it had just done to the *R. W. Griffin*, her captain, and her crew. The vessel was just far enough away and in a swirl of mist and sea spray that we could not get any identifying numbers or a name. By the time we were able to haul back, secure our gear, and steam to the approximate site of the collision, we found nothing of any consequence. There was some debris, some shattered bits of the dories that had been secured to the top of the wheelhouse, and little else.

There had been no eyewitness to the collision, no survivors to present a description of the ramming, and certainly no statement of responsibility from the Russian ship. Nothing. Eight men were lost that night, one of them a friend. All were fellow fishermen, and what happened to them could have happened to any of us.

What still stands out in my mind are the words of the assistant secretary of the interior, whom I had talked to in Washington: "One missile base in Iceland . . . is worth more than all of the fisheries in this country . . . " But was it worth the lives of eight men?

A MOTHER'S SON

"This looks like a weather breeder to me," Norman said to no one in particular as he scanned the sky off to the sou'west. The wind had been increasing over the last few hours, but the day looked beautiful to me. The glass (barometer) was holding around thirty millibars and there was an increasing ground swell, but the surface of the water was as smooth as silk. What wind was blowing was as warm as that of any spring day you would expect offshore. But Norm's comment stuck in my mind. He had been at sea longer than Noah, and he was seldom wrong in his prognostications. Time would tell.

Looking for butterfish, we had set out the first tow in fifty fathoms and were working offshore on the 1820 line on the loran chart. We were making short tows, mostly half-hour shots, looking for a sign.

When we saw fish running out of the top square of the net's mouth as it broke the surface, we knew that we were not far off from the mother lode.

Using the 1820 loran line as a base, we worked in different directions, in a pattern much like the spokes of a wheel, until we found the main body of fish. At least that's what it was supposed to be like, but when you are dragging a net over the sea bottom, you never know what will happen next. That's the excitement of the business. You can't be sure what's going to come over the rail of the boat.

We were fishing short-handed this trip; the crew consisted of Norm, Paul, and me. Still, that was nearly a century of experience combined. This time of the year we seldom ran into really heavy weather unexpectedly, our gear was in good shape, we had just hauled the boat out, and everything above and below the waterline was now shipshape. Still, I'm a firm believer in Murphy's Law—"If anything can go wrong, it will." I can't believe an Irishman came up with that saying, but I've seen it happen too many times to dismiss it.

The day was half along, Norm and I were below having a mug-up, and Paul was in the wheelhouse when we felt a surge that nearly upended our coffee. We were on the way topside before Paul gave a blast on the air horn. He had pulled back on the throttle but had kept the boat in gear to maintain headway against the rising sou'west swell. I ran aft to look at the tow wires, and sure enough, they had come together. We were into something that I was sure would be unpleasant.

I shifted positions with Paul, and he went aft to release the hookup around the tow wires. Before he got there, however, I called out to him to hold up. "I'm going back over the wires to see if I can tow the net off the hang." This maneuver might be successful if we had not dragged the object too far into the net. As I brought the boat around and we doubled back over the wires, they began to hum as our speed through the water increased. I slowed down when I figured we were about at the location where we had snagged, and we began to tow in the opposite direction. It didn't take long before I knew that we were not successful in breaking free.

This being the case, I turned back to a southerly direction, as I wanted to be downwind when we hauled the net.

We would usually make a slow circle until the doors came up and would continue the loop until the ground wires running from the doors to the wire legs that allowed the headrope to rise were up into the gallus blocks. Then, when we were downwind of the net, we would throw the engine out of gear and lay off the net as we hauled it up to the side of the boat.

Things were handled differently when we were hung up—especially when we had no idea what we had snagged. We would maneuver ever so slightly. This time, as soon as we came downwind of the net, I threw the boat out of gear; there was no steaming around in a circle until the net was up to the rail. Things began to jingle as soon as we began to haul the wires in. This was no bag of mud or a rock or even a piece of some wreck that would begin to break apart if it had been in the water for very long. Whatever we had snagged was extremely heavy. The winch began to chatter, the rigging was vibrating with the strain, and I was happy that we had recently put new tow wires onto the drums of the winch, as the worn cable that we had taken off would never have held this weight. Of course, after this strain, we would have to remeasure and re-mark the wires since they would be stretched too unevenly to fish right.

"Let's go real slow, fellers. I want to get back as much of the net as we can," I called, my head stuck out of the starboard wheelhouse window.

Ever so slowly we inched the net nearer to the surface.

What will it be this time? I wondered. *Another torpedo ready to explode, or did we run into some old fishing boat that went down without its position being recorded on the charts?* There was no sense in worrying; whatever the object was, we would have to deal with it soon enough.

"Paul, stand by the rail and let me know as soon as you can see the net and what it's hung into. But keep your head inside the cable running between the gallus frames." Running from the forward gallus to the after one, this cable had a block and sheave at each end that were used for the ropes leading to the quarter sections of the net. By using these lines, we could haul just the mouth of the net while the wings were left overboard.

Within a few minutes Paul yelled out to Norm, who was manning the winch, "Slow it down; I can see the gear coming. The net is still spread out pretty wide, so whatever's in it isn't too far down inside."

Norm continued to winch the net ever closer to the surface.

Again Paul called out: "Oh, my God, Cap! You've got to see this. Hold it, Norm, we don't want it any higher." I jumped down to the deck from behind the after drum of the winch after putting the brakes on tightly. Leaning over the side, I couldn't believe my eyes.

Lying just below the surface and almost directly under the *Dorothy & Betty II* was an airplane—a whole, complete airplane—a Grumman TBF fighter bomber of World War II vintage. I knew exactly what kind of plane it was, as I used to see them flying around the Salt Pond area throughout the war.

"Norm, I want to bring that plane up as close as we can without its hitting the hull," I said. "Paul, climb in beside Norm, and run the after winch at the same speed that he does. But go slowly, both of you!"

As my crewmen engaged the drums of the winch, everything on board began to vibrate, hum, screech, and moan with the strain.

"Okay! That's far enough! I've seen enough!" I cried. The bright sunshine filtering below the surface of the water lit up everything that needed to be seen.

"Secure the winch and take a look," I said as I turned toward Norm and Paul. When they reached the rail and looked overboard, they took in the grisly scene hanging just below us. Not only had we hooked into a plane, but the crew was still aboard. We could clearly see the pilot, secured in the forward seat. His hatch was open and pushed back as it would normally be when he was landing. He was wearing his leather helmet and goggles, and he was still strapped into his parachute.

The navigator, who also served as bombardier and rear gunner, was sitting just behind the pilot, fully dressed in his gear. They both looked as though they were about to touch down at Quonset Point, the likely home base for this crew.

Paul was staring down when he said, "Man, both of them bought the farm on this one."

I replied, "No, Paul, all three men bought it. There was a belly gunner on the bottom of the plane, and no doubt he died first as they crashed. He had nowhere to go. He watched it happen from the best, or worst, vantage point there was."

I could just imagine that gunner talking with the pilot as they came closer and closer to the water, until that moment when they struck with so much speed and force that it might well have torn the belly turret right out of the plane. We couldn't confirm this, but I knew from the type of plane that there had been a third man aboard.

As we discussed our situation, I became aware that scuds were increasing overhead and the wind was freshening; it would not be long before the seas began to make up, and that would put us in real trouble. There was enough strain on the gear as it was without a surge to contend with.

We were unable to see any numbers on the fuselage nor any names; we knew nothing other than the kind of plane we had hauled up and where it was located. (At the first opportunity I had made a quick notation on the chart.) One thing I did observe was that we had dragged our net right over the cowling, engine, and cockpit before hitting the wings. Since we were towing in a southerly direction, that meant that the crew was likely heading north, probably back to Quonset. The plane appeared to be intact, so the pilot had been outstanding in his flying ability as he ditched the plane. Moreover, the Grumman had not been shot down, as we could see no evidence of battle on the fuselage or wings. I believe they just ran out of fuel before they could make Block Island or some other landfall.

What a tragedy. More than a decade had passed since this crew of airmen had left Quonset Point Naval Air Station in Rhode Island, completed their mission of searching for enemy submarines, and—apparently—headed back to their base. With their mission accomplished and only a few minutes of air time left, they would soon have been back in the loving embraces of family, friends, and comrades had it not been for the lack of fuel.

As we looked down at this very sad end of the lives of three young men, I realized that each of them was some mother's son, some child's father, some woman's husband, and it saddened me to leave them in the watery grave where they had been resting this long time. But we had no choice. True to Norman's earlier observation, the day had truly become a weather breeder. The once-high glass began to fall, the sky had darkened, and the wind began to back around to the southeast, increasing considerably. Heavy rain would shortly be upon us. The sea began to heave its weight around, and our small craft had to take quick action or we, too, could be in serious trouble.

"Let's take the bolt cutters and cut the forward legs to the net," I said. "That should allow the plane to drop quickly and tear its way out of the body of the twine, hopefully leaving enough for us to mend."

Once that decision was made, things happened quickly and like clockwork. In just a few minutes, the forward wing of the net thrashed loose from the rail, submerging immediately, and the plane fell quickly to the seabed that had been its resting place since the end of the war.

Some might wonder why I made the decision that I did. I could have called the Coast Guard, given them our position, and asked for assistance. But ours was not a life-threatening situation, and it would have taken days for someone in the higher chain of command to authorize a vessel big enough to pick the plane out of the water. And due to worsening weather conditions, we would not be able to hold the TBF at or near the surface. It had been all we could do to raise it on a fair day.

And I could have given a description of the plane, telling of its occupants. But what good would that have done? Three families somewhere in the United States had already wept and prayed and mourned until the pain was bearable. Would it have been a kindness to rekindle that pain? Nothing would have changed. No one was alive. Those airmen had been resting in that watery grave awaiting some greater call on some future day. Nothing that I might have tried to do would have improved their situation.

By the time we brought the trawl doors aboard, chocked them in by the gallus frames, retrieved our very tattered net, and salvaged what few fish remained in the cod end, we were in the midst of a southeast breeze of thirty knots or more and ever-higher seas. The only good thing that came from the weather change was a fair wind and a following sea, making our trip home smooth and quick.

Yes sir, you never know what the next day will bring when you set your net over the side to tow the bottom of the sea.

MAKING ICE

There are plenty of life-threatening events that can happen to you at sea: You can get run down by a tanker or other larger ship. An uncontrollable fire can break out on board. Your boat can strike a submerged object while steaming at night. Or, you can tow up live ordnance that explodes under your vessel. And the list goes on.

Here, however, I'm talking about something so gradual and has with a cumulative effect that you can be in really serious trouble before the danger becomes apparent. That danger is "making ice," when the sea that is breaking over the bulwarks and the spray that is blowing across the deck into the rigging, gallus frames, and nets, begins to turn to ice. When I was a skipper, this was the most insidious threat faced by the crew of a New England fishing trawler. Just listen:

We had been fishing for butterfish just inside the continental shelf in about ninety fathoms of water. It was January, but we were working on deck in shirtsleeves because the temperature was so warm. The water flowing around us came up from the Gulf Stream off of the Florida coast. We were working on the butterfish as they traveled with that warm water.

Fishing had been very good and we had iced down around fifty thousand pounds of "butters." We hauled in the net after our final tow, cleaned it, and secured it tightly to the bulwarks. The drag doors were chained down in their slots between the bulwarks and the gallus frames. The decks were cleared of fish. Baskets, shovels, and all small equipment were stowed below in the fish hold. Hatches were secured with heavy steel bars and then covered with canvas, which was tied down to prevent seawater from flooding the hold. The whips and block and falls were lashed down tightly. Even the deck hose was coiled up and tied, so it wouldn't wash overboard while we were steaming home.

The cook prepared the last big meal we would be able to enjoy, as we would be steaming to the north'ard and the seas would be coming over our starboard quarter. This meant that all the solid green water would hit the heavy steel gallus frames and be busted up, thus minimizing its impact on the companionway house and the smaller "doghouse" forward. The companionway provided entrance to the fo'c'sle via a built-in ladder that allowed access belowdecks. The doghouse had two small hatches that allowed more air below as needed. When we were steaming in rough weather, both hatchways were closed to keep seawater from soaking those below.

Cooking on a fishing boat is difficult enough on the best of days, let alone on the bad ones. So we usually took it easy on the cook when we had to steam to wind'ard. The meal had been prepared and the cook had come aft to relieve me at the wheel as the rest of us went below to mug-up as quickly as possible. Then we steamed slowly until the cook had a chance to clean up the fo'c'sle from the meal and to wash the pots, pans, and dishes. As soon as he had everything secured, we would pick up speed and the crew would begin standing watches of two hours each.

On bad days, I would sleep in the bunk in the wheelhouse. That way I could quickly take the wheel while the man going off watch crawled forward over the deck, hanging onto the lifeline we had rigged between the main mast forward and the winch just ahead of the wheelhouse. When he reached the whaleback, he would go below to wake his relief. That would go on until we came near land, or some other situation arose and needed my attention; then I would take the wheel to bring her in or to get through any difficulty that might have arisen.

When everything was secure and all the crew but the first man who would take a watch were in the wheelhouse with me, I increased the throttle to the "rusty notch," or flank speed. When carrying the amount of fish that we had aboard, the *Dorothy & Betty II* was like a floating island when going to wind'ard. She would rise slowly on each sea and then settle rather gently into the next. Water would cascade over the bow, break into foam as it struck the rigid steelwork bracing the gallus, then run off the decks through the scuppers along each bulwark.

We had been steaming toward home for about six hours and the wind was increasing from the nor'east, causing more and more water to break over the bow and run down the deck. I had slept for a few hours since we began our homeward run, so I told Paul, the last man on watch, not to wake his successor, as I would "hold her" until we made the breakwater at Point Judith. With the increasing seas and men having to cross the deck in all of that water, it was getting too risky to keep up the shift change.

Paul checked the engine room and reported that all was well. I slowed the diesel and watched him as he waded across the deck, entered the companionway, and closed the doors. The deck lights showed that the doghouse was closed and secured from below, so I again began to increase engine speed and settled down for a run of six hours or longer, depending on what happened with the weather and the sea conditions. While the overhead lights were on, I looked closely around the deck and up into the rigging; everything was secure and clear. The air temperature was still quite warm—probably thirty-five or forty degrees—so other than a headwind and rising seas, everything seemed shipshape.

Because of the weather change, I turned the radar on and set it at the thirty-mile range. The screen was clear. I engaged the autopilot, turned on the overhead light, and decided to finish a book that I had started on the way out from the Point. I checked the depth sounder, and we were showing fifty-five fathoms.

Since there would be no changes in the watch, I had no reason to turn on the deck lights for some time—perhaps a couple of hours. I kept myself busy by periodically checking the engine room, keeping an eye on the radar screen, working through some candy bars that I had stowed beneath my bunk back in the chart room, and reading my book. The sounding machine indicated that we were now well inside the fifty-fathom curve.

I had not turned on the radio since we had started for home. Had I, we would not have been sliding into the situation that was progressively developing on deck. The last weather forecast I had heard was the 11:20 P.M. report for the waters from Eastport, Maine, to Block Island, Rhode Island. What they had projected was just what we were experiencing: "increasing northeast winds backing around to the west sometime tomorrow." The announcer said nothing about rapidly dropping temperatures.

I didn't know it, but the "insidious threat" mentioned earlier was slowly encasing the *Dorothy & Betty II* in a cocoon of ever-thickening ice. The line of frigid air lay along the fifty-fathom curve, right at the edge of the Gulf Stream. As spray hit the rigging and seas washed across the deck, ice was forming quickly.

No sooner had I turned the radio to the channel used by local fishermen, than I caught the words of one skipper who had just hauled back on the edge of Deep Hole, east of Block Island. He said that when he hoisted his net up to the blocks and then dropped it to get another hold on it, "The twine was standing up straight as a tree and not falling because it's so cold." I immediately snapped on my deck lights and saw only a dull glare; the wheelhouse windows were covered in ice. I tried to open the window on the leeward side without success. Next, I tried to open the doors on either side of the wheelhouse. They, too, were stuck fast with ice. I was about to pull the lanyard on the air horn to get the crew on deck, when I realized that they, too, were probably locked in. If they woke up to the horn with no prior knowledge of the weather on deck, they might think that we were about to get run down and would find themselves unable to escape the fo'c'sle. Since they were happily asleep and unable to help, and because we were in no immediate danger, I decided to work things out by myself.

One option was to turn around and head back offshore, into deep and warmer water. But turning broadside to the heavy seas when I didn't know how much ice had made up was too risky.

Doing so, on the other hand, would allow time for the ice to melt. Not knowing how thick it was, I had no idea how long that would take, and if someone woke up below and tried to get out, that would create a problem. On the other hand, if I continued steaming due north, toward Point Judith, our ice load would increase, and if we became top-heavy, we could roll over and sink.

I decided to change course and head for Old Harbor on Block Island. That would be more westerly in direction and would reduce the amount of spray coming over the bow and blowing upward into the rigging. I checked the radar screen, and thankfully it was high enough above the wheelhouse that the ice had not jammed the rotating antenna. There were no blips on the screen, and the island showed up nicely.

As we neared Block Island, I changed the radar range down to twelve miles, and the breakwater at Old Harbor showed up nice and clear. There was no traffic near the harbor nor headed on a collision course with us. I thought about calling the Coast Guard to advise them of our problem, but the station was in New Harbor, on the west side of the island. Before they could get to us I would be either in Old Harbor or on the bottom.

Steaming ever closer to the island and being mindful to watch for nearby traffic, I finally switched the radar onto the one-and-one-half-mile range. Not only did the breakwater jump up crystal clear, but the buoys almost leapt off the screen. I could see the short breakwater wall inside the harbor, and I could even make out the boats that were lying alongside the main wharf inshore.

Now, I want you to know that I had never done anything like what I was about to do, and I wasn't even sure how accurate this radar really was. I had only owned the *Dorothy & Betty II* for a short time, and had never worked with radar before buying her. In fact, she was about the first boat in Galilee to have radar. Well, I was about to find out just how good it was.

As I approached the end of the outside wall of riprap that protected the harbor, I slowed the engine very gradually, as I did not want to startle anyone below. Peering into the radar, I then turned inside the wall and headed about southwest to parallel it, maintaining just enough speed to give me headway. I was nearing the end of the inside wall when I began to yank on the air horn, ripping off four long blasts and then repeating that series. This was a distress call known by all fishermen. As I did this, I threw the engine out of gear and allowed the *Dorothy & Betty II* to drift slowly toward the boats that showed up on my screen. When it looked as if I was about to make contact, I jammed the engine into reverse and increased the throttle. When my headway had been all but halted, I pulled the engine out of gear again and let my remaining headway carry us alongside the boats tied to the dock. Before I could feel any contact, I could hear men jumping aboard, and they yelled that they were tying us down. Then I could hear them calling to the crew below, in the fo'c'sle, and I heard the sound of pounding on the ice.

In short order they had cleared the wheelhouse and the companionway. I met my crew on deck and quickly explained what had happened. Then we collectively thanked the crowd of fishermen that had been so quick to grasp what our problem was and respond appropriately.

Later that day the wind came around to the west, and the temperature rose above freezing, allowing us to head back to the Point.

Did I make the right decision in heading for the island instead of for warmer water offshore? I don't know for sure. It worked out well, but might not have. It was a calculated risk, much like those I took every time we went to sea. But you can be assured that on the next trip to the fifty-fathom curve, I had a man in the wheelhouse with me, and I listened continually to the weather station on the radio.

Oh, yes, and I frequently checked for ice.

AN ARMFUL OF DANGER

The *Mary Alice*, owned and captained by Fred Gamache out of Galilee, was not the most seaworthy of fishing vessels in the harbor, but she was a step up for me. I had set my sights on the highliners in the port, but you had to pay your dues by advancing one category of boat at a time, graduating from day boats to small trip boats to large trip boats and then to highliners such as the *North Wind* or the *Princess* or the *Rhode Island*.

Day boats, as their name implies, leave the harbor at or before daylight, steam for an hour or two, fish a variety of grounds, then return to unload their catch by late afternoon. Trip boats may come back into harbor at the end of the day, but they usually remain offshore at least overnight, and more than likely for several days. The prime factors involved in the length of time they stay out are the weather, the abundance of fish, or the price of the catch being landed.

The *Mary Alice* was one of those vessels that vacillated between being a day boat and a trip boat. Sometimes the skipper himself was not sure what he was doing or where he was going. The day I signed on board we were to head south of Martha's Vineyard after swordfish. Fred had fully rigged the *Mary Alice* for that kind of fishing, outfitting her with a stand, or pulpit, about twelve feet long, several harpoons, warps, and kegs, and a dory free of junk ready to launch in the event that the skipper put an iron into a broadbill. Yet, bottom trawls were still tied down along the bulwarks, and drag doors were secured to the gallus frames.

When readying their boats for swordfish, most skippers cleaned all gear off the deck except what was needed for that species. Every nonessential item was looked on as a hazard—not only to the crew but also to the smooth functioning of the harpooning gear. After a fish was ironed from the stand up forward, the warp running from the "lily" (or harpoon dart) to the keg, which sat on deck (sometimes as far aft as the transom) would literally fly out of the basket that held the coils. If anything, such as a trawl or drag doors, was snagged by the warp, the sudden pull of the swordfish would likely yank the lily out of its flesh. Since Fred had left all of the drag gear on deck, I didn't think that he was very serious about swordfishing.

My observations were correct. Fred knew that I had been rather successful as a striker on other vessels; he also knew that I was looking for a site on a harpoon boat and would not be interested in dragging at this particular time.

We had been steaming for about three hours, and I was the only hand aloft watching for the identifying fins of a swordfish. I knew that no one could sit on the portside crosstree of the mast, as there wasn't one. The wood had rotted out, and the timbers had fallen to the deck sometime before. Yet none of the crew were on top of the wheelhouse or on the ratlines looking for fish, and that didn't set well with me.

The engine slowed down and Fred turned the boat so she would lie before the wind. When that turn was nearly complete, he threw the engine out of gear and we laid to. When I climbed down to the deck, all hands began to take the lines off the nets that had been stowed and to raise the drag doors in preparation for swinging them overboard. I asked Fred why we had stopped short of our destination, and he said that he had changed his mind and wanted to drag for yellowtail flounder. End of discussion.

After several tows to the westward, toward Block Island, without great success, we were coming to the end of daylight. Since our net was heavily chained and was dragging deeper than the fish would go, we began to pick up a considerable quantity of sea scallops. When the net was hauled for the last time that day, we were told to set the doors on deck, that we were going back to Point Judith.

Soon the scallops were shucked, they and the fish were iced belowdecks, and any minor damage that the net had incurred during the day was repaired. I went into the pilothouse to stand my watch at the wheel. Fred's brother Jim had set up the helm in the fashion of the old fishing schooners, with the helmsman standing to the side or slightly in front of it. All other fishing boats that I had been aboard installed the wheel in the forward bulkhead of the wheelhouse, and the helmsman would stand behind it. The latter arrangement was much more comfortable, especially on long watches. At any rate, I held closely to the compass course given me. As we steamed along, Fred said that we were going to change gear when we arrived in Point Judith; we were going scalloping. Man! Three changes in one day.

Upon arriving in Galilee I made a few quick calls to see if any boats in port were going swordfishing. None were. The only ones so rigged had already left, and were on or near the fishing grounds. So I decided that I might just as well stay on the *Mary Alice* for a while. Scalloping was better than being without a site. At least I could make some money while waiting for the swordfish fleet to return.

Just as I had guessed, it took several days to change over from dragging to scalloping. We had to remove all the drag gear, and Fred had to buy a couple of dredge frames. Then we had to build the bags that would be dragged along the seabed. These consisted of case-hardened steel rings hooked together with unbelievably hard steel links.

Once a bag was complete and secured to the dredge frame, the combined weight of the rig was well over a thousand pounds. From this point on, all hands were at risk of being crushed should they get between the scallop dredge and any part of the boat. With the dredge swinging around due to wave action, serious injury could result in a matter of seconds, especially while crew members were working on a slippery deck.

Now that the *Mary Alice* was fully rigged, we needed to take on fuel, water, ice, burlap scallop bags, wires to tie these bags closed, the special tool used to turn the wire, shucking knives, cutting benches to dump the scallops on for the shucking operation, and— most important—food, lots of food. The best thing about this upcoming trip was Henry Beebe's cooking, which was always outstanding.

With all preparations complete, we steamed out of the Point and headed toward the scallop grounds. Although disappointed that we were not going swordfishing, I was looking forward to an uneventful yet profitable trip. Profitable: Yes. Uneventful: far from it.

The weather was great, the fishing was bountiful, and we would be heading home in a day or two. One concern that I had was the amount of seawater the *Mary Alice* was taking on. We had two pumps working steadily in the engine room to keep ahead of the flow there; and the five-inch deck pump had to be manned frequently to keep the bilges in the fo'c'sle and the fish hold clear. Henry, the cook, said that the *Mary Alice* was a strip-built boat. In other words, she was constructed upside down on a frame, and each plank was nailed to the strip below and not into a rib or frame as on all other boats at the time. The ribs were added only after the basic hull was complete. This method created a very limber hull. Sea action would loosen the hull, and water would seep into the bilges at a rapid pace. Still, as long as the engine kept running, we were able to keep ahead of the deluge.

Things went smoothly until the first tow on our last day of fishing. The sou'west breeze was picking up as the sun rose when we began to haul the forward scallop dredge up to the block in the gallus frame. Before we could hook the whip onto the dredge, the increased rolling of the vessel caused it to slam into the wooden sheathing on the side of the boat.

As the dredge swung outboard with each wave, the man standing by the forward gallus frame could see something hanging in the dredge framework. Because of the dredge's rapid movement, he couldn't identify just what the object was. He called out to the man on the winch to be extra careful as he hauled the dredge aboard. But the skipper stuck his head out of the wheelhouse and told the winch man to get the dredge aboard as fast as he could. He was concerned about damage to his relatively fragile strip-built boat.

As one man released the drum holding the cable used for towing the dredge, the other deckhand threw three turns of whip rope onto the winch head and began hoisting. Just as the frame mouth came up to the top of the bulwark, we could clearly see what sort of object was hanging and banging around in the framework of the dredge. It was a hedgehog—the explosive device found in one type of depth charge.

A depth charge, at least the kind most commonly dragged up by commercial fishermen in our area, looked much like a fifty-five-gallon metal drum. Inside the drum were a half-dozen tubelike explosive devices, each measuring about four feet long and twelve inches in diameter.

With the rolling sea and slippery deck, it was going to be difficult to retrieve the weapon, but the job had to be done—and quickly. It was clear that the device was only hanging by some fins that had become jammed in the dredge framework, and any motion could release it, with tragic results. Henry and I leaned overboard enough to reach the hedgehog, and with one upward pull released it from the dredge.

Our feet were slipping this way and that as we walked aft and laid it on deck alongside the wheelhouse. While I held it in place, Henry found some planking to use as chocking, and within a few minutes the bomb was secure. While we were working, the skipper called the Coast Guard at Point Judith. He described our find, advised them that we would complete our fishing for the day, and told them that we should be in port by 7:00 P.M. He asked that the navy provide a demolition team to remove the hedgehog while we unloaded our catch at the Fishermen's Co-op in Galilee.

We continued fishing, with one eye on our work and the other on the innocent-looking means of destruction. Just imagine: That thing lying alongside the wheelhouse was capable of destroying a submarine five times or more our size and killing all hands. If it went off aboard the *Mary Alice,* nothing would be found of our boat and crew. With such thoughts running through our minds, we kept on hauling and setting the gear.

Upon arriving at the Co-op, we slid into the open berth on the south side. Fred headed our bow in toward shore, which meant that when the navy boys arrived, they would have to carry the weapon the entire length of the boat, climbing over our fishing equipment and deck gear, then moving onto the Co-op dock, and heading for the ramp where they would have parked their truck. This was not a good scenario.

By the time we had the *Mary Alice* tied up, the navy had arrived. We had been expecting a contingent of troops to relieve us of our unwelcome visitor. But who showed up? One very salty, weather-beaten chief petty officer with a well-chewed cigar stuck in his mouth. Quaking in his shadow was an emaciated, pale, young sailor.

"Well, Skipper, where is that #@%^&* that you are so concerned about?" bellowed the chief.

Fred, who was equally salty, said, "The !%*&&! thing is back by the wheelhouse and you &$%**(@ well better get it off my boat before it explodes!"

"Well, swabby," said the chief to the sailor, "you heard the captain! Get your $^#! down on deck and get hold of that bomb."

"Bomb?" stammered the sailor. "I didn't know we were going to pick up a bomb."

"Listen to me, swabby, that !#@$^%*& isn't what you want to be afraid off. I'm the one you better worry about. Now move it!"

With that sweet invitation, the sailor jumped down to our deck and, landing on some ice, flew into the air backside first, banging his head on a deck checker.

"Get your *^&)##@! butt off the deck and get moving!" hollered the chief, who then climbed down the ladder to our deck and headed aft with a rolling gait. As he got to the wheelhouse Fred handed him a crowbar and hammer. "Here you go, Chief. You might need these."

The Coast Guardsman ignored his offer, and approaching the hedgehog, he reached down and yanked the device free of its blocking. "Now hear this, seaman. You will lift this bomb and carry it midships to the Co-op boom. You will hold it until I tell you to let go. Do you understand, sailor?

"Yes, sir," quaked the seaman. He bent over and tried to pick up the hedgehog without success.

"Get out of the way, you !$@^#%! Let a man at it," said the chief, and in one easy motion he lifted the bomb, turned around, and dumped it into the outstretched arms of the young sailor. "Now move it!"

The burdened young man stumbled his way forward, his feet going this way and that as he stepped on fish slime and ice. Sweat poured off his forehead, and his face was chalk-white with fear. For a moment he appeared to fall backward, but the encouraging words of the chief empowered him.

"Listen up, swabby, if you drop that $*%%(#) bomb you will be mine to nurture for the balance of your enlistment. Now, I don't think that you would enjoy that one little bit. Would you, sailor?"

"No, sir. I'll try harder."

"Try! You will more than try, sailor, you will do! Do you hear me?" bellowed the chief.

Without another word, the sailor stood up and walked as straight as an arrow to the portion of the deck below where the boom from the Co-op had swung out and dropped a whip line to lift the bomb off the deck.

The chief strapped it around the weapon and signaled the man on the winch to lift it up and onto the co-op dock.

"Well, swabby, you did #!%&**$@ good. I was sure you could. You know, if you had dropped that thing it would have cleaned out most of this harbor and everyone in it." With a big grin on his face and his well-chewed cigar nearly gone, he turned toward Fred. "Well, Cap, the *%&&#@! navy saved you again." He turned and climbed up the ladder secured to the Co-op bulkhead and disappeared.

I might have been disappointed about missing out on swordfishing, but the trip with Captain Gamache was certainly exciting.

SWORDFISH AND OTHER WONDERS

I have often said that few sights seen by men ashore can equal those observed by men at sea. Without exception, they cause one to feel a sense of awe, and this is especially so of the Creator's larger specimens.

One such is the swordfish, a magnificent creature of wondrous power. One fisherman from Block Island, Rhode Island, who had been very successful at harpooning swordfish, began losing those he had struck. Once ironed, the fish would dive, the warp would run out, and before the keg was pulled overboard the line would go slack. When the warp was hauled in, the end would be cut as cleanly as if someone had sliced it with a knife. After this happened a few times, the skipper quit fishing and headed inshore to a shipyard.

When his vessel was hauled out, the yard crew found a swordfish sword driven through the keel, which was nine inches thick, built of solid oak. Just think about that. The sword had passed completely through and then had broken off, leaving enough sharp edges on both sides to slice a warp in one stroke. Can you imagine how much power it took for a fish to accomplish that feat?

One time on the *North Wind* we were fishing the peak of Brown's Ledge, just south of Nova Scotia. It was a gorgeous day—flat calm—and fishing was good. Jerry Adams, our skipper and an outstanding harpooner, was out on the stand, and I was aloft, steering from the topmast, when a fish came to the surface within striking distance of the stand.

With a single fluid motion, Jerry swung around with the harpoon in his hands and ironed the fish as sweet as could be. Now, we were steaming at about eight knots in one direction, and that broadbill was swimming across our bow at probably four or five knots. It had just cleared us when the iron hit, and in a flash the fish turned sharply back toward the boat, leaped out of the water, and punctured the bulwarks, falling back into the water after breaking off part of its sword. Jumping again, that broadbill struck the top plank of the hull with its broken sword and drove it into the planking. Again the fish broke off its sword and fell back into the water. Unbelievably, the broadbill drove itself upward a third time, striking the hull yet again and driving what remained of its sword into the second plank of the hull, breaking it off again, and collapsing once into the water, where it lay stunned. Our doryman quickly retrieved it.

Measuring, we discovered that the swordfish's first strike hit the *North Wind* about eight feet above the waterline, and each of the following strikes was about a foot below that one. To give you some idea of how fast all this occurred, make a fist and then pound it as fast as you can into your other hand three times. All the while, I was aloft, looking down upon this amazing demonstration of speed and strength. Later, as we sat in the fo'c'sle having supper, Jerry said, "If swordfish had any degree of intelligence, there wouldn't be a wooden boat afloat." I believe him.

It always amazed me how differently individual swordfish reacted after being struck with the harpoon. You would expect them to dive as quickly as possible, and that frequently happened. Yet I have seen more than one fish give a quick kick of its tail, throwing up a plume of white water that drove it a yard or two away from the boat, and then quietly swim on the surface in the general direction that it had been heading before being struck.

One such fish slowed down just as Jerry saw that the warp—which, thankfully, happened to be brand new—was running out at a moderate speed but had not been properly secured to its keg. As Jerry hollered to someone to catch the tag end, it went overboard. Being aloft, I could see the new, dry warp just lying on the surface, so I hauled the steering rope hard to port, kicking the stern closer to the fish. Art Raposa, who was standing on deck, grabbed a pitchfork and, leaning over the rail, was able to retrieve the warp. He brought the tag end on board, then ran forward and tied it to the keg, which he threw overboard just as the ironed fish decided to dive.

One thing that a doryman does not want to see is a fish that goes wild when hit. It will typically sound and then turn right around and breach, thrashing its sword one way and its tail another, whipping the air in an apparent effort to strike whatever hit it. Once a broadbill does this, it is not likely to stop acting wild until hauled and killed.

I remember a trip when I was number-one doryman aboard the *North*

Wind, meaning that I would go out to tend the first fish ironed. My brother-in-law, Roger Beaudet, was number-two doryman. Jerry had just harpooned a beautiful, big swordfish, and it was a wild one. It began to buck and thrash and dive and breach, again and again. Roger began to tease me about the good time I was facing in tending this "crazy fish." In the meantime, Jerry signaled the helmsman to turn around and head toward the keg that was attached to this bucking bronco of the sea.

As soon as we came close enough for the doryman to leave the mother ship, Jerry called out, "Roger, this is your fish. Get in the dory." Well, you've never seen a man's expression change as quickly as Roger's did. He jumped into the dory, pulled the keg aboard, and began to haul the warp, spinning and bobbing with the fish's wild gyrations. Getting it to the rail took considerable time, but he eventually succeeded. Roger never openly teased me about wild swordfish again.

Once a doryman picks up the keg that has been put out on an ironed fish, he hauls in the warp, keeping a constant strain on it as he does so. If the warp suddenly goes limp, he will throw a handful of coils as far overboard and as quickly as possible. The change in tension means either that the dart has pulled out or, worse, that the swordfish has turned around in its dive and is heading back to the surface at tremendous speed, backtracking the warp leading from its injured back. When a fish like that breaches, it often has death in its eyes. Many a doryman who failed to take the precautionary measure of tossing coils of warp off to one side of the dory has paid with his life when the fish followed the line to its source and attacked the boat with its sword.

My brother, John, a doryman of some renown, was once hauling a fish while observing a nearby dory from another port. The man was standing in the bow and hauling hard when the fish turned, followed the line, and came up through the bottom of the dory, driving its sword right into the man's leg and up into his stomach, all the while twisting and turning. By the time my brother could get over to help, the man was dead. John told me that story when I was about to make a trip as a doryman, and I never forgot it. Since then, I have had a number of fish breach right through the coils of warp that I had thrown overboard.

Sharks

Many, if not most, people react to the effects of a storm brewing—you know, the aching joints, the dull pains, and an overall malaise—and I believe that creatures of the sea are affected in a similar way. My entire crew once observed such a change in the personality of blue sharks, a usually docile creature.

Consider the time we were fishing just south of Martha's Vineyard in about thirty fathoms of water. The day was clear and bright, with just a light southeast breeze. We were in hopes of catching butterfish, and the first three tows had shown a few large ones in the top square of the net. We were obviously near the mother lode.

Each time we emptied the net and threw trash fish overboard as we cleared the deck, we observed blue sharks circling just a few yards away. They were like puppies trying to snatch a favorite toy. They would nudge the net, roll over onto their backs as they made a leisurely dive, and then slowly make another circle, eating any fish they encountered in their easy revolutions.

While we were hauling the fourth tow, I noticed a line of dark clouds east of us. They had appeared suddenly and had an ominous look to them, so I turned on the radar to see just how big they were. The image on the screen suggested that a line squall was making up. As soon as the doors were hooked and the legs and ground cables were wound in, we hauled the quarter ropes, bringing the mouth of the net up to the rail. As the crew dropped the mouth and began to whip the body of the net in, I ran into the wheelhouse to look again at the barometer, because each minute that went by caused me more concern over the storm mass that was moving quickly toward us. The glass was dropping rapidly.

"Tie everything down, fellers. We're in for a stiff breeze," I called out.

The onboard portion of the net was secured, the doghouse and wheelhouse doors were closed, and the fish hatch was secured. Just then, the wind struck. It blew so hard and so quickly that I thought we were in a tornado or a waterspout. Water was whipped into the air, and the net lying overboard was no longer hanging straight down, as it normally would. Instead, it swung off the side of the hull like a boat ramp used on the beach.

The once-docile blue sharks that had been ambling along beside us, showing no cognitive perception at all, now turned into raging sea creatures, their eyes and insatiable hunger directed at the men on the deck of the *Dorothy & Betty II.*

In a combined century of experience at sea, not one of us had witnessed what was now taking place. Those sharks left their natural habitat, the ocean, and with extreme gyrations, worked themselves up the ramp-like body of twine, snapping their jaws and heading for us. The men on deck ran over to the port side, and I jumped into the wheelhouse. The line squall was passing us by, and a quick look at the glass showed it beginning to rise as fast as it had dropped. Within minutes the seas calmed, the net began to settle, and the once-crazed sharks again began quietly circling, looking for any butterfish that had been squeezed out of the body of twine.

We finished hauling the net and ended up with a large bag of butterfish on the deck, all to our delight. Things had returned to normal and we made one more tow. But we didn't stop thinking about those sharks.

Whales

On one trip aboard the *North Wind,* while fishing for swordfish near the Peak of Brown's Bank, we witnessed the most thrilling sight that I have ever experienced. We had been looking for finning swordfish, but on this day without success. Suddenly one of the men aloft in the topmast called out, "Breacher!" indicating that he'd seen something leap and fall back with a splash.

"Where away?" asked the skipper.

"To the nor'west, on the horizon," came the reply.

Jerry called up to me to change course to the bearing given. Thinking that what had been observed was a swordfish, we headed that way. It is generally agreed that when swordfish breach, they are either trying to shake loose the large worms that commonly bore into their flesh or, as a group, they are about to change grounds. In the event of the latter, the fish don't stick long after breaching begins, so our long jog might be in vain. Still, some sign is better than no sign. As we progressed northward, we continued to look underwater ahead of and around the boat, as the larger fish that show up late in the season on Brown's often fail to horn out, preferring to travel just a few feet under the surface.

We saw that the breaching was continuing. It was unusual for swordfish to carry on for such a long period, but we were happy to see it. As time passed—something like half an hour—we realized that what we had been seeing were not swordfish a short way off but something far different and much farther away. As we approached the location of the breachers, we were all struck silent by the wondrous sight before us. Two humpback whales were diving to the depths of the ocean, then driving full speed for the surface, clearing it by their full height and rolling belly to belly while doing so.

These leviathans would appear to stand on their flukes before they pulled apart, flipped onto their backs, and struck the water with their huge pectoral fins. What we had been seeing for all those miles as we steamed into their vicinity was the spray from the fins as the whales hit the water, throwing it fifty feet or more into the air.

Fishermen are notorious for foul language, but as we circled this awesome sight, not a word of any kind came out of the mouths of the six men aloft, and the crew on deck was silent, too.

Scallops

One night as we were steaming just south of the Nantucket Lightship and approaching the western edge of Georges Bank, I stepped out of the wheelhouse and leaned over the side to check on the deck hose that was pumping water from the fish hold. I wanted to switch the valve as soon as the bilge in that compartment had been emptied.

As I looked down, something flashed by the side of the boat, then another and another. Since it was dark, my vision was limited. I quickly climbed into the wheelhouse and flicked on the deck lights. On each side of the *Dorothy & Betty II*, a body of sea scallops was swimming right along with us! I then turned on our spotlight, and everywhere in its brilliant beam, which extended outward for a quarter mile or more, were swimming shellfish. Now, we were steaming at nine and a half knots, and we watched that body of scallops for a half-hour before they thinned out and then disappeared. I have no idea how long we had been running through them before they caught my attention, but we had seen tens of thousands of these creatures.

Only a couple of the crewmen below were awake enough to come up on deck when I called down into the fo'c'sle. Neither had ever witnessed or even heard of such a phenomenon before. My best guess is that the scallops were simply taking advantage of a fair tide to change grounds, although I was unaware that they in fact did so. Since no one has offered any better answer, I'll stick with my conclusion.

Seabirds

Seabirds have always intrigued me. During the summer months, two species were ever present when we were fishing. I didn't know the proper name for them back then, but fishermen always called them Mother Carey's chickens and hags (or sea hags). Today I realize that they were stormy petrels and common shearwaters, respectively. Especially when we were swordfishing and steaming over the ocean for miles each day, both of these species would accompany us constantly.

Mother Carey's chickens were about the size of a mourning dove or a little smaller, were black with a white rump, and had a rounded tail. Considering the number of miles they had to fly each day, they had relatively short wings. They would constantly hover over the ocean, dipping repeatedly to the surface, and were apparently feeding all the while, although I could never see what they were eating. I never saw them rest on the water, drifting debris, or the pods of seaweed that we would sometimes run across. I know that they must have stopped flying at some point, but I never observed it in all my years at sea.

In spite of their name, hags were a more attractive seabird. Their backs were slate-gray to black, and their bellies and underwings were white to light-gray. (My wife says that I am color-blind, so the descriptions of both these birds are suspect.) They were larger than the Mother Carey's chickens, too, with longer, pointed wings.

Hags were very graceful as they glided and swooped and skimmed food from the surface. They covered a much greater area in their quest for nourishment than did the Mother Carey's chickens, and they had the grace of the eagles that today fly over my inland home.

Like the chickens, they never seemed to stop flying. To my knowledge they have never been seen resting on the deck or in the rigging of any fishing boat.

I always thought of the gannet as the clown of the seabirds, perhaps in the same class as the puffin, although I never saw puffins on Georges Bank or the Northern Edge, both areas that we fished extensively. Most of my observations of gannets took place far offshore, perhaps one hundred or one hundred and fifty miles out, along the edge of the Gully, or continental shelf.

As we hauled up our net while in pursuit of the silvery butterfish, thousands of the smallest ones would slip out of the twine and dart toward the depths from whence they came. While this was taking place, hundreds of gannets would circle overhead until they spotted the butterfish "heading south." They would then gracefully fold their wings close to their sides and dive at a blurring speed, hitting the water's surface with tremendous impact. If the fish were out of reach, the birds would open their wings underwater and "swim" after them, usually with success. Catching a butterfish in its mouth, the gannet would again tuck its wings into its body and then turn toward the surface. Air within their bodies would catapult them two or three feet above the waves. Once stabilized in a sitting position, the bird would swallow its meal, stretching its long neck and fluffing its feathers before looking underwater for another course. If the gannet saw fish nearby, it would simply duck beneath the surface and begin swimming. If needed, it would take off, fly to the necessary height, and begin the initial procedure over again.

Although graceful divers and swimmers, the gannets had a comical appearance, with a relatively large, white body; a long, gangling neck; and a head that looked like the Concorde with its nose lifted in flight. Black rings nearly circled their eyes but didn't quite make the connection. They were a sight to behold and brightened many a somber, rainy, stormy day.

Whale Sharks

Another sight that puzzled us for a number of years had first been seen while we were steaming north at swordfishing speed—about eight knots.

We were on the northern edge of Georges Bank and our destination was the peak of Brown's. The swordfish were close to leaving the Eastern Seaboard and would be heading out to the Flemish Cap and then to the Mediterranean. We would not see them again until they showed up south of Block Island the following July. So, we were anxious to top off our catch and head home.

Jerry Adams was sitting in the stand as it rose with each crest and then dipped down until the water was just inches from his feet. We were all waiting for him to get dunked in a really big sea. At least that would bring a little excitement into what had been a boring, hot afternoon.

There were six men aloft, and we were supposed to be straining our eyes for the ever-elusive broadbill. But, a man can only stare into the bright sun reflecting off the ocean's surface for so long without getting sleepy—and possibly falling out of the hoop holding him safely in place, which meant he would drop sixty feet to the deck. So we tried every mental game we knew, finally drifting back to our favorite target: the captain.

It was during these mental high jinks that we heard from the man sitting "on the ball," the topmost position in the mast, whose sole job was to look underwater for fish that the others might miss while searching the surface for fins. He called out, "Forward port quarter, a hundred yards!"

Here at last was some excitement. We all began to look underwater in the direction given. A large gray mass was heading toward us just off our port side. There was no head to be seen, no fins visible above or below the water, and no tail to identify the species. The creature was not "blowing," so we quickly ruled out a whale of some kind. The only markings that we could see were large white circles about the size of dinner plates running in rows along its sides. This mystery beast was moving, but it seemed to undulate in place rather than swim.

Just think about this: The *North Wind* measured fifty-six feet, and as we came abreast of the great "fish," its back alone was longer than the boat. If you allowed another ten or twenty feet for the head that was apparently hidden in the deep, and the same or even more for its tail, we were looking at something over a hundred feet long. No one aboard had any idea as to what we were seeing, and it kept us guessing for the rest of the trip.

Several years later, my wife and I were watching a movie about Thor Heyerdahl sailing a raft from the Mediterranean to South America. Sure enough, one scene showed the self-same creature bumping up against the raft, nearly upsetting it. The narrator called it a whale shark, one of the largest creatures in the sea. "At last," as Inspector Cluso would say, "The case is sol-ved."

Fish Fire

Phosphoros is the ancient Greek word for "light bearing," and it survives today in "phosphorus," the word used by fishermen everywhere to describe the eerie glow they often see in the water at night. Back in the 1800s, though, mackerel seiners called this phenomenon "fish fire," and when a big school stirred up the phosphorus, they would say, "The water fires well tonight." Sending men aloft, into the rigging of their schooners, the old-timers would steam around their favorite fishing grounds on calm, moonless nights until the lookouts spotted a large area of luminescence in the ocean. They would then lower their boats and surround the illuminated school of mackerel with the purse seine, a huge curtain of twine that could be closed at the bottom to trap the fish.

These days—or, rather, nights—phosphorescence in the sea is simply regarded as a gift from the Creator for the delight of fishermen everywhere. When you are steaming along on a dark night and "the water fires well," the sight is breathtaking. This is especially true when a school of porpoises decides to accompany the boat, diving under the bow one moment, streaking ahead the next, then crisscrossing just feet ahead of the bow. It's a wondrous event that one never tires of, and it ends all too soon.

Turtles

Most little boys delight in finding a box turtle and cradling it in their small, chubby hands. These docile creatures are often put into pockets and taken to school for show-and-tell.

Most of these same youngsters are unaware that turtles grow much, much larger. An old-timer harpooned a big leatherback turtle off Block Island one day, and he and a deckhand struggled to get it aboard their boat. Upon arriving in Galilee, they trucked their catch to a local mortician and had it embalmed. They then bought an old hearse with glass sides. Next they loaded the turtle into the hearse, painted WORLD'S LARGEST TURTLE on the sides, and visited the local fairs, charging folks twenty-five cents to look at it. Another Barnum was born.

Although the public was amazed at the sight, that turtle was not in any way the largest seen by fishermen. I've seen leatherbacks as big as a household sofa. U.S. fishermen have never bothered to catch them and bring them to port as there is no domestic market, and it takes too much time and effort.

Sunfish

Another large creature seldom seen by anyone but a seaman is the ocean sunfish, or *Mola mola,* which—appropriately enough—is Latin for "millstone." With no exaggeration, I've seen them as big as a king-sized bed. They lie on their sides at the surface with only a large, flabby dorsal fin flopping slowly as they propel themselves along. The sunfish's tail is an odd-looking, squared-off stub, and its body is solid as a rock, as anyone foolish enough to try harpooning one has discovered. Incredibly, these ungainly creatures can and do breach, leaping and falling back into the water with the splash of a barn door dropped out of a second-story window.

Lilliputians

While fishing the continental shelf out to a hundred and fifty miles offshore and depths up to a hundred fathoms of water, we would see creatures so small that they would easily fit into the palm of the hand. Most were without names: shrimps, crabs, sponges, and flat, weedlike creatures, most of them as white as snow. I believe that they often came from much greater depths—over the "edge," where the bottom quickly dropped five hundred to a thousand fathoms (or more) in a very short distance.

As a fisherman, I found every one of these wonders beautiful, whether it was small or large, enchanting or fearsome. And I looked on all of them as creations of the "fifth day," when God got to see that "it was good."

THREE-FREEZE CIDER

It had been a long, hard week, and we all were exhausted. Because of unstable weather and the prospects of a hurricane making its way up the coast, Ted Dykstra, our skipper, had decided to work Deep Hole, Sou'west Ground, and the area west of Block Island in the days remaining before the expected storm. The fishing had been productive enough, but all of these grounds meant making day trips: up and out no later than 3:00 A.M., and then short tows all day long, "playing buffalo" (your head down and your butt up) on deck for twelve hours or more. It was far different from fishing offshore or down east, where we made longer tows and the catch was much "cleaner" (meaning there were very few unmarketable fish), giving us a chance to rest a bit or have a "mug-up" between tows.

Deep Hole, in particular, was a place that demanded the close attention of the man in the wheelhouse, and the fellers on deck had to be ready to haul back quickly when we started to get into trouble. Rocks, wrecks, mud, and sundry other delights would be the consequences of even one moment of inattention on the part of the captain, who was watching the depth sounder, the loran, and—in fog—the radar. Staying on course along the sharp edge of the hole required some fancy turns, and one misstep meant we'd be in "net-mending city" for a day or two. But if you could avoid the hangs, you were usually well rewarded.

The interesting thing about the Hole was its assortment of fish. On most grounds you worked on one species of fish—perhaps yellowtail flounder—and there would be little else. But at the Hole you would find blackback flounder, whiting, a few butterfish, perhaps a cod or two, hake, and even more. If you had a good fish finder, then you could distinguish each body of fish as you went over them while towing.

In fact, one fisherman from Snug Harbor, Clint Babcock, got so good at this that he could tell you what species were in the Hole, what fish were in which layer as they swam, and about how many of each species there were, by weight. It was uncanny, but he proved himself accurate time after time. I, on the other hand, could hardly see any fish on the bottom—no matter how many there were. As long as we filled our net, that was the important thing.

I mention all this to explain why we were so tired by the end of the aforementioned week. We left the Fishermen's Co-op at Galilee after taking out our fish, topping off the fuel and water tanks, filling the hold with the necessary amount of ice, and then steaming across the harbor to the West Side (the Jerusalem side), where the *Anna Grace* tied up. She was fifty-seven feet long and had a schooner hull designed by Howard Chapelle, the famous naval architect. She was not rigged as a schooner, though, as two big masts would have taken up too much deck room aboard a vessel intended for dragging. The *Anna Grace* was seakindly—comfortable in any sea condition that we found ourselves in. And that turned out to be a blessing on many a trip.

Don Morse and I shook down the net (clearing any fish, crabs, etc.), then hauled it up with the boom, allowing us to hose down the deck and put things in order for another day. Meanwhile, Ted was in the wheelhouse listening to the latest weather report.

Don and I were about to do some mending along the mouth of the trawl when Ted came out on deck to tell us that we could put off the mending until the next day, as the hurricane was picking up speed and would be hitting us twenty-four hours sooner than expected.

It didn't take us long to drop the mending needles, grab our sea bags, and heave them onto the dock.

"We can catch up on the repairs tomorrow while I change the oil," Ted said. "Let's head home. We can sleep in a little tomorrow morning. I'll see you around seven."

Don and I just grunted, picked up our gear, and dragged ourselves up the dock toward our pickups.

At 7:00 A.M. sharp Ted's pickup swung into the parking area on the State Pier. Earlier, Don and I had met in town at the coffee shop and had decided to head down to Jerusalem in one truck, so I dropped mine off at Flanagan's Shell Station to be serviced. We brought along some extra coffee and donuts for Ted, so we wouldn't have to make breakfast aboard the boat.

All of us sat in the wheelhouse drinking our coffee while Ted tuned the radio to the weather station. We really didn't need to hear the forecast, as the glass was dropping rapidly, and the sky was darkening with heavy clouds. While driving along the beach on our way to the boat, we had seen the waves increasing in size, and the water was taking on that strange, heavy look common to bad weather.

The news was as expected: the storm was now a full-blown hurricane, with winds of seventy-five to ninety miles per hour. It appeared to be increasing in both size and forward speed as it neared shore. The hurricane was expected to hit Point Judith late that night, or no later than the next morning.

We set about doing the work Ted had scheduled. Once that was done we hauled up some heavy hawsers from belowdecks and began to lay them out on deck in the order of their being used. The dock that we laid to was just on the north side of the State Pier at Jerusalem, and that huge structure provided us with considerable protection from any seas that might crash over the walls of the breakwater.

It would also shield us from drifting wreckage or boats that might tear away from their moorings. In addition, there were heavy pilings abreast of our bow and stern, on the offshore side of the *Anna Grace*, and we ran some of the hawsers out to them. This would prevent the boat from being slammed into the dock, especially as the hurricane moved by and the winds shifted to the east and then the northeast.

When all the lines were secure and Ted was happy with the way that the *Anna Grace* was lying, we concentrated on the nets, extra lines, and any loose deck gear such as shovels, baskets, and fish picks. These we stowed in the fish hold and then battened down the hatches. We put out the fire in the fo'c'sle, secured everything that might get tossed around, and put a lock on the icebox to keep it from popping open during the thrashing that was expected.

When everything was shipshape and "Bristol Fashion," as the saying goes, we headed up the dock. "Let's come back tonight at six and we'll ride out the storm on board," Ted said as he was getting into his truck. "If we have to, we can head out into deeper water. We'll come down in my pickup, and my wife can drive it back home."

When I got home I checked out the yard and house to be sure that everything was secure enough to withstand the expected high winds.

Ted came by at 5:30 P.M. after picking up Don, and we headed for the boat.

As we drove toward Jerusalem we ran into a military roadblock where the road divides at the Gooseberry Road, leading to Snug Harbor. "We're heading down to my dragger," said Ted to the National Guard sergeant as he strode up to the truck. "My wife will be coming right back, but we plan on staying aboard," he continued with a smile. I guess he thought he might charm the man wearing a sidearm and carrying a rifle. But neither the sergeant nor the six other troopers were any happier about seeing us there than we were about seeing them, so Ted's smile didn't move him at all.

"If you go down to your boat now, you won't be able to come back out, as we have orders to turn back any traffic—effective as soon as you leave. Do you understand what I'm saying?" the sergeant asked calmly but very firmly.

"Yes, sir, I do," replied Ted. "We'll be staying aboard until the storm moves by."

"The storm has increased in its movement and intensity. It will be coming over the breakwater and into the harbor by midnight. After this, we are not going to allow anyone beyond this point," replied the soldier. "Good luck," were his final words, and they left no doubt that we wouldn't be returning from the *Anna Grace* that night.

I thought to myself, *I sure hope Ted has this planned well,* but I had my doubts.

"My father and Bill are taking the *Anna D.* up the pond, and John is moving the *David D.*, as well," Ted said as we continued on our way. "But we don't have to go up there. We'll be secure at the dock. Besides, the *Anna Grace* draws nine and a half feet of water, and there isn't any sheltered spot that we could get into."

We arrived at the boat and brought our gear aboard, Ted said good-bye to Grace, and she headed back to town. I looked around the village in Jerusalem, and not a soul was to be seen. All the small boats were gone, as their skippers had the same idea as Ted's father and brothers. They were up on the east side of Salt Pond and north of Great Island. There they would be sheltered from the high winds.

I was sure that my grandfather had done the same with his trap boat, the *Olive*. All the families, children, pets, automobiles, and trucks had skedaddled up to the high ground. It was just Ted, Don, and I, all by our lonesome.

We climbed aboard the *Anna Grace*, started the engine, checked the bilge pumps, checked the lines, and added a few more where we thought they might be needed. Don went down into the fo'c'sle and started the fire in the cookstove. At least we would have some hot chow and coffee, so things wouldn't be all that bad. *Not right away, at least,* I thought mournfully.

As it turns out, this was the first named hurricane to strike Point Judith head-on. The storm was named Carol, and she was far from being a lady. Her strength and violence were exceeded only by those of the September 1938 hurricane. Carol had sustained winds of ninety miles per hour, with gusts of one hundred ten to one hundred fifteen. This was the "breeze" that we were to ride out. Well, we tried. As soon as we finished supper, we washed the dishes and secured them in the lockers. Don then put the fire out, as we did not want to have flames running wild in the fo'c'sle during the height of the storm. We would have enough to contend with.

The tide was rising rapidly, as was the wave action inside the harbor. What with the driving rain and mist, we were unable to see the breakers as they began to top the seawall. In fact, we could hardly make anything out across the harbor, in Galilee. All the streetlights and the lights on the buildings along the bulkhead were out, leaving everything in the growing darkness. What we could see before night set in was not a pretty sight. Fishing boats and yachts alike were rising well above the docks they were tied to, and many settled down on the pilings that had been put there to protect them, punching holes in their hulls. Other boats were adrift and were being blown up against the docks at the north end of the harbor. Because of the flood tide and violent wave action, many of the smaller craft had washed up over the docks and were being smashed onto the surface of the parking lot and into the walls of the fish-processing plant.

Now we had our own situation to worry about. With our deck lights on, we watched as the *Anna Grace* rose higher and higher toward the tops of the pilings that were on our east side. The heavy dock lines were keeping us from being hammered into the inshore side of the dock, but within moments, the hawsers were going to work their way up and over the pilings.

Ted put the boat in gear and tried to jog slowly in place, with the idea that when the dock lines lost their purchase, we might be able to stay in the lee of the State Pier. So far, it had been protecting us from derelict boats and other debris that was being swept into Salt Pond. In just a few moments the lines went slack, but we were, in fact, able to hold our position. Don and I left the wheelhouse to haul in all the hawsers and put them down in the engine room so they wouldn't wash overboard and get into the propeller. The wind, which was driving the rain right straight across the deck, was mixed with sand and was acting like a sandblaster. I hunched down with my back to the wind until we got the lines secured, and then I dove into the wheelhouse. What with all three of us in raingear, boots, and sou'westers, there wasn't too much room to move around, but anything was better than being out in that wind. Ted did a great job of holding us in between the pilings and behind the pier, and it looked as if we might make it after all.

That was not to be. Ted looked down at the gauges and moaned, "The engine is heating up. I think that the seacock is getting plugged up with weed and other debris. If the diesel gets much hotter I'll have to shut her down." The engine's temperature continued to increase by the minute.

"Get your life jackets, quickly!" Ted barked. As soon as I put mine on, I took hold of the wheel while Ted donned his. "If I don't shut her down, we'll lose the engine," he said, and as he was speaking he hit the kill switch.

We were now in a life-or-death situation: The engine had stopped, we had lost all control of our vessel, and we were at the mercy of the hurricane, the waters around us being driven by raging winds. Oddly enough, I thought of an old salt's lament, "I'd rather be on land making water than on water making land." I guess that I was becoming fully cognizant of our dangerous situation and just wanted to be home with my family.

The *Anna Grace* was rapidly being pushed up into Salt Pond, and since there was nothing but sand flats for a mile or more to the west and east of us, the area would soon be inundated by the docks, boats, and buildings of Snug Harbor. The wind whipped up oceanlike seas all around us, creating a heavy spray that made it nearly impossible to tell just where we were heading. It wasn't until we came to the flooded houses on Strawberry Point, where the water from Potter's Pond came roiling out into the main body of normally placid Salt Pond, that I knew we were just a few hundred yards from my grandfather's dock at Snug Harbor. I also knew that a few hundred yards farther north lay a deadly ridge of rocks that could tear the bottom out of our boat should we strike them at this speed. Whatever happened was now strictly a matter of chance. We had absolutely no control, and since it was too nerve-wracking to conjure up all the horrible ways our wild ride could end, I put them out of my mind.

I said to Ted, "This flow of water is bouncing us off the shore by Cap's docks and Knight's Shipyard. I think we may well end up beached on or near High Point."

"I do, too," came his reply. "It couldn't be a better location as it'll put us near the deep-water channel when the time comes to get her off. But we better get ready to jump ship as quickly as possible, wherever we land. Get to the highest ground you can find, and stay there."

The words were no sooner out of Ted's mouth than the *Anna Grace* began to list so hard to port that we were thrown into the side of the wheelhouse. She was beginning to ground out on the shallowing water beneath her. She bumped again and again, and we began to wonder if she would ever stop. Finally, she did, and Ted yelled to Don, "Open the starboard door and get out! Quickly, now."

The door was heavy, and as hard as Don tried, he couldn't lift it. Then both of us heaved against it, and the door swung straight up before flopping back with a thud against the outside of the wheelhouse. Don climbed out first, then turned to haul me out. In turn, both of us helped Ted climb out. We then shut the door to keep the driving rain out of the wheelhouse and, we hoped, out of the electronics. Slowly we crawled toward the bow, and one by one we dropped down the steeply sloped deck, all the while hanging onto the winch and then the rigging to keep us from dropping into the rushing water that was boiling up from under the hull and flashing by into the storm-roiled channel just a few feet away. Of the three of us, Don was by far the strongest. Planting his feet in the sand, he leaned into the bulwarks and grabbed Ted, then me. He held onto us until we found some stable gravel beneath our feet. Ted and I then yanked him to safety.

We ran a few yards up the beach, stopped, and turned toward the boat. Only then did we grasp the reality of what had just happened: We had survived a five-mile ride through a turbulent hurricane in a disabled dragger, had been cast like dice onto the sand by the storm, and had lived to tell about it.

The most amazing thing about all this was that the peninsula of land on which we had grounded was High Point, owned by Ted's brother Jake. We climbed the hill to his house and went in until the hurricane abated. I was so exhausted that I didn't even take my raingear and boots off before falling asleep on the floor. The next morning we awoke and walked about a quarter-mile to Ted's parents' house. Mrs. Dykstra gave us a huge breakfast, then Ted borrowed his dad's car and we headed for Wakefield. No phone or electrical service had been restored, so our wives had no idea where we were or what might have happened.

As it turned out we couldn't get home, because the same National Guard sergeant we had met just before the storm hit was still standing guard at the junction of the Jerusalem and Snug Harbor roads. This time he was adamant about our not getting through. He said that his men had to secure the area and prevent looting of the properties along the beach.

We were so tired and wound up that Ted never balked about being stopped. He turned the car around and took us back to Jake's house. "We'll try again tomorrow," he said quietly. "I'll see about getting a pass from the military governor at the town hall in the morning, then I'll give you a call. If the phones aren't back on line, I'll stop by." I got out and climbed the stairs to our apartment. Gloria had seen the kids through the hurricane all by herself, and they were safe. What a trooper! She never said a word about the storm once she saw that I was all right.

We made out far better than many others. By the time Hurricane Carol decided to move on, it had left nineteen people dead in its wake. Three thousand eight hundred homes were destroyed, and there was more than ninety million dollars in property damage. Carol had struck Rhode Island on August 31, 1954, a day that will live, as they say, "in infamy." At least in my memory.

It was noon before Ted stopped by with a permit that would allow us down to High Point to inspect the boat. By the time we got there, Ted's father had brought the *Anna D.* back from safe harbor up in the pond, and John had followed suit with his boat, the *David D.* They had ridden out the storm without incident. Well, nearly. John had sailed across some high ground during the height of the hurricane, passing between a tree and a summer cottage and dragging a deep trench across the well-manicured lawn with his keel. He was certain that the homeowners would be upset, so he planned to repair the damage as soon as he could get over there.

By the time we got through the roadblock and down to High Point, there must have been a hundred sightseers on the peninsula. Atop of this outcropping of rocks and gravel was a carpet of loam and moss, and crowning this gorgeous spot was the beautiful home belonging to Jake Dykstra. Extending to the house from the mainland was a very narrow driveway made up of rocks, shells, and other debris. It was commonly inundated by any particularly high tide, but a pickup could always cross it without difficulty. Now, that driveway allowed access to the most unusual sight of the *Anna Grace* lying over on her port side, with the bow pointing due east. I walked around her feeling embarrassment for the naked, awkward state in which she sat on this spit of land, in full view of the residents of Snug Harbor. I privately assured her that we would take care of her as quickly as possible.

Just then, Ted ran up, excited as all get-out. "She's in perfect shape, no damage, but the really good news is that another hurricane is making its way up the coast as we speak."

"That's good news?" I replied.

"Yes—don't you see? We can use the storm tide to refloat the boat; it shouldn't be more than a couple of weeks at the outside." It seems that the hurricane he was talking about was making up off Florida and had already been given the name Edna.

I had my reservations about the idea, but Ted was making a little bit of sense. The part that I didn't particularly like was the "we." I countered, "You mean that we can climb back aboard this boat that is now lying on her side with the scuppers in the water, and we can wait for a storm tide that might raise her enough to float her off the beach?"

"Yes," he replied, expressing full confidence that the wind and the tide and this now-inert hull (with an engine that may or may not have been damaged and might or might not run) would all cooperate in the relaunching of the *Anna Grace*. Ted was so positive that I began to believe him, and I conceded that his idea was a brilliant one. Don later agreed, although reluctantly.

As the second hurricane made its way up the coast, we carried out the first phase of Ted's plan by moving our bedding and groceries up into Jake's house, some sixty feet above the *Anna Grace* and the water in which we were hoping to refloat her. We intended to sleep in the house until we heard the storm reaching its peak; then we would go aboard the boat and put the final phase of Ted's plan into action. We left our trucks up on high ground on the mainland after offloading them and walked back across the narrow causeway and up the hill to the house, which would allow us the best view of the approaching hurricane.

Ted lit a fire in the living room's beautiful fireplace. Jake's house was graced with windows facing all corners of Salt Pond and Galilee to the south. Hardwood floors and exposed wooden beams added to the comfortable ambience of the rooms. I began to cook up a meal on the gas stove, and Don took it upon himself to reconnoiter the attic and then the basement. He was apparently hoping to find a bottle of booze that might have been overlooked by the homeowners when they vacated the premises before the arrival of the hurricane.

It wasn't long before Don climbed the stairs from the cellar carrying several bottles. Without a word he set them down, scurried right back down into the basement, and quickly made a second trip. He repeated this circuit several more times. I was busy cooking and didn't investigate the treasure that Don had found. He finally completed his trips to the cellar, but instead of resting he began to arrange the bottles in some semblance of order on top of a large cupboard. All the while, he never said a word, but he began wheezing a bit—not from climbing the stairs several times, but more out of anticipation.

I eventually called to Ted and Don that supper was ready, and when we sat down to eat Don spoke up in an excited voice and said, "You'll never guess what I found in the cellar." Before Ted or I could respond, he explained: "I found more than a dozen bottles of Foster Browning's cider. Can you imagine, *Foster Browning's*, of all things." Obviously, Don was a connoisseur of cider and had a particular fondness for that produced by Foster Browning. I knew of Foster, as I went to school with his son Steve, but I had thought he was strictly a farmer. Apparently, the man had other gifts.

After the supper dishes were cleared and washed and put away, Don very proudly moved his find to the dining-room table and began to organize the bottles into four distinct groups. When we asked why, he pointed to some old glass milk bottles from Foster's dairy farm and said, "These are bottles of regular cider." Then Don indicated a group of old, chipped Warwick Club ginger ale bottles, explaining, "These here are one-freeze cider." Next, he showed off some well-used Narragansett Lager Beer bottles and said, "These are two-freeze cider." Finally, Don gestured to a group of newer Moxie bottles with their bright labels still intact and announced, "These here, these have three-freeze cider in them." This meant that their contents had been aged for at least three weeks, a process that began only after the cider's temperature first dropped below thirty-two degrees.

Don was almost drooling when he identified the collection of three-freeze bottles. When I asked him about the nomenclature of this last grouping, he seemed to gasp in astonishment. "Don't you know anything about cider?" Without a pause, Don went on to describe the process of cider making as practiced by Foster Browning. I'm sure that all of us know generally how cider is produced, but Foster apparently took that process to greater heights.

"When the weather turns cold," explained Don, "Foster pours his cider into large galvanized washtubs, leaving them outside through the night. The next morning, before daylight, he skims off the ice and bottles some of the cider. He then labels it 'One-Freeze.' He continues this process after the second cold night and then after the third. By the time Foster skims the ice off of the final batch, it's near the potency of white lightning, and these are the best-selling bottles in South Kingstown and beyond," Don concluded almost breathlessly.

As he was finishing his explanation, he picked up three glasses and opened one of the one-freeze bottles. As he was pouring, I asked him about the black sludge that was streaking the liquid. He lifted the bottle to the light and said, "That's nothing but a few fruit flies," and kept on pouring.

"Look, Don," I responded, "I'd like to try this elixir of life, but I don't do bugs."

"Oh, don't worry; I can fix that," he said. And he stopped pouring, went to a cupboard, and found a small dishtowel. He grasped it tightly over the neck of the bottle and poured the cider into a different glass. Sure enough, the fluid came out ruby red and clean as a whistle. Of course, when Don took the towel away from the cider bottle, nine million and three very dead, well-preserved fruit flies lay on the soggy cloth with their feet toward heaven. But the cider was now as clear as vintage gasoline,

Swept inland by the high winds and floodwaters of Hurricane Carol, Ted Dykstra's dragger Anna Grace *was left high and dry after the storm.*

with much the same fragrance. Fortunately none of us were smokers, and Ted had let the fire in the fireplace go out. If the fumes had hit an open flame, I'm sure the conflagration on the top of High Point would have been visible all the way to Martha's Vineyard.

Never being the coward in a group, I held my breath, put the glass to my lips, took a big gulp, and swallowed, throwing all caution to the winds. As the liquid hit my stomach and created a belch beyond my control, I turned away from Ted and Don so as not to singe their faces.

"Very nice," I barely managed, in a voice that sounded more like Don

The only way to refloat the fifty-seven-foot vessel was to build a temporary cradle and railway beneath her, then slide the hull into water deep enough to float her.

Corleone's as he ordered a hit on some Irishman in South Boston. Those were my last words for a while, as my vocal cords needed to heal.

Ted, a Methodist at heart, declined a drink and decided he would turn in. I declined another belt and headed for my sleeping bag. I signed to Don to turn the lights off before going to bed. He nodded an affirmation and turned to the task at hand: drinking the Foster Browning collection into oblivion.

I woke at daylight, and as I lay there, I could not hear anything—no rain, no wind—and I could see no clouds, just the dawning of a beautiful, clear day. (We would later find out that the hurricane we were counting on to refloat the *Anna Grace* had fizzled when its "eye" had split prior to reaching the Rhode Island coast. The storm's maximum winds were just forty to fifty miles per hour.)

I woke Ted and we went downstairs to the kitchen area to find Don asleep with his head on the table. Not on his arms, but flat out, with his forehead on the wood. He failed to hear my calls or Ted's hollering, and he was not stirred by my nudging or Ted's violent shaking. Don was in Foster Browning's seventh heaven. Or was it his eighth purgatory? Obviously, Don wouldn't be able to tell us for a while.

As I looked around, I saw several empty bottles with a residue of dead flies at the bottom. A few other bottles sported dishtowels, each with a smudging of flies. Then there were still others where there wasn't a single fly in the bottle or wiped on a towel. It was all very interesting, but the answers to our questions as to what had happened would have to wait until Don's revival—if he revived. In total disdain, Ted left the house and went down to the boat. I joined him shortly.

After making a careful inspection, Ted decided that we would have to assemble a crew of men to build up some blocking under the hull, then construct a railway of sorts—more like a greased slide to ease the progress of the boat into deeper water. He would start on that project as soon as he could get back home. (Ken Van Duzer of Peace Dale, a house mover of some renown, met with him before the day was over.) Ted asked me to stay with Don and to get him home after we cleaned Jake's house of our clutter and the odd collection of bottles.

Don stirred a couple of hours later, and when he lifted his head from the table, I saw that deeply embedded in his forehead was the impression of the metal opener he had used to get Foster's three-freeze beverage out of those Moxie bottles. As soon as I thought that Don's mind had cleared a bit, I asked him about the various fruit-fly remains: some in small mounds on the table, some still in the bottles, and some on the towels. I inquired about the squeaky clean bottles, too.

He squeezed his head, then held it under the cold-water faucet until I thought he would drown. At last, he sat down and pondered the events of the night before. Finally he said, "At first I strained the flies with those soggy dishtowels. That worked quite well for a while. Then I decided that I could drink the cider faster if I just strained the flies as I drank, spitting the bodies onto the table. That worked well for a while, but then I said 'What the hell' and I just drank the cider, flies and all. That worked best of all, but the next thing I remember was you waking me up with all the noise you made gathering up the empties. Boy, I don't want to see a bottle of Foster's three-freeze ever again."

At that moment, I didn't know how long it would take to get the *Anna Grace* back in the water and fishing once more. But I did know that Don was not going to be able to "play buffalo" on deck for a long time, what with the hangover he was suffering. In fact, I don't think that he could even have handled fishing down east or offshore with those long tows and easy cleanups. I just hoped that he would be able to pull himself together enough to face Ted. As it turned out, Don paid most dearly for his indiscretion, as Ted assigned him every dirty detail that came up for a number of months afterward.

LIVE ORDNANCE

"Let's haul back, fellers, the wires are coming together," I called out to the crew, who was busy cleaning up the deck from the last tow. "Moe, you and Paul put the fish below. Norm, I'll let the wires go and then you can start bringing the gear in."

It had been an ideal day to fish. We were sou'west of Block Island in about twenty fathoms of water, looking for yellowtail flounder. Our first tow after daylight brought in about six bushel, and I was making each tow to the east'ard of where we first set out. I was hoping to find an even bigger bunch of fish, although six bushel was certainly worth working on, as the price was as high as we had seen it in a month or more.

The weather could not have been better. In the vernacular of the fishermen in Point Judith, the sea was "flat-ass calm," and there wasn't a cloud in the sky, which was rare on such an early-spring day. We'd had such a string of bad weather during the last part of the winter that it had taken its toll on men and boats; a day like this one was a keeper. There were no other boats around, so if we found good fishing, we stood a good chance of working on the flounder for a while before being overrun by the boys from Stonington, Connecticut. They generally fished this area. In fact, that was the strangest thing of all—to be this close to land and not have a small fleet of day boats under our bow. At the time, though, I was happy with the situation.

We were on our fourth tow, and the third one had brought in nine bushel of "tails" (yellowtail flounder), so I guessed that we were heading in the right direction.

Neither my logbook nor the chart showed any wrecks or soft bottom in the area, so I was surprised when the tow wires started to come together. If you snag a wreck or if you tow into a soft bottom, the boat will generally surge almost to a stop, and you can feel it throughout the vessel. This had been a relatively slow-developing incident, which usually meant that we'd hung up on some object and were towing it along the bottom. At any rate, this was not good news, as anything caught up in the net would mean a lot of damage to the twine.

Well, I thought, *we can haul the rock or other object out of the way, bend on another net, and keep fishing.* I did not want to lose any more time than necessary, not with this good weather and an increase of fish with each successive tow. Paul climbed up out of the hold. He had just iced down the fish that Moe had dumped into the open deck plate.

"Get that plate secure. The doors are nearly up," I called to Moe.

Paul had moved forward and was ready to manhandle the first drag door when it came out of the water, hooking it into the steel gallus frame. We used two seven-hundred-fifty-pound rectangular wooden doors, each measuring about six feet long, to hold the net open as we towed it along the ocean floor. At the lower edge of each door was a heavy, case-hardened steel "shoe" that took the wear and abuse as the door dragged and bounced along the bottom.

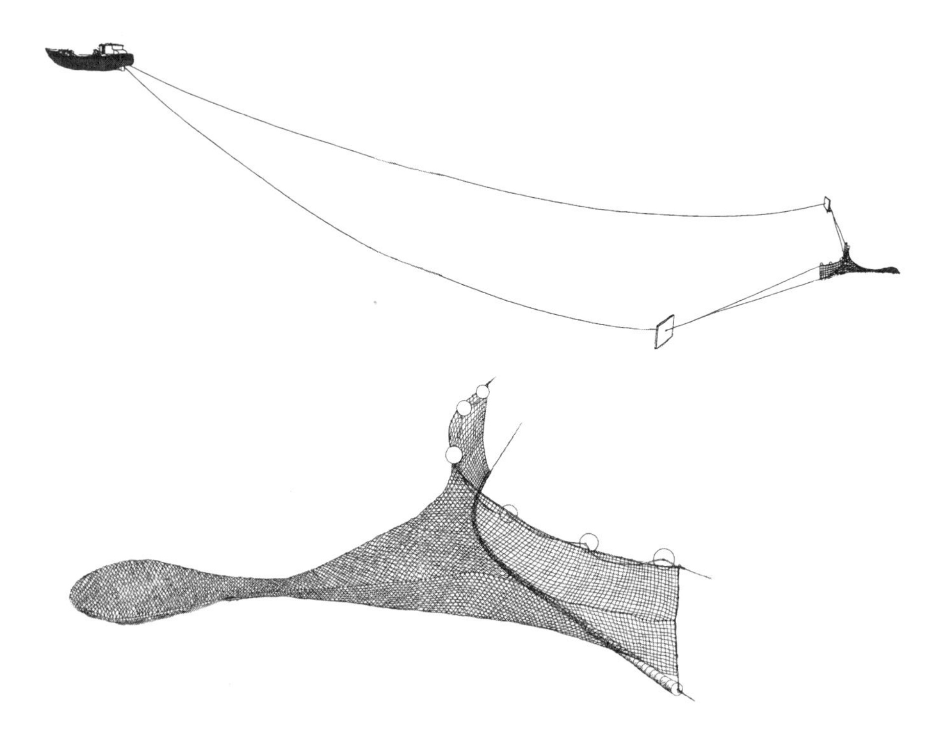

Two views of a typical "otter" trawl show the net on the bottom, towed by steel cables and spread by a pair of trawl doors. In the close-up are the floats that raise the top of the net and the rollers and weights that maintain its contact with the bottom. Note the baglike cod end, where most of the fish end up.

In the meantime, Moe had come aft to help Norm run the winch. It took two men to control the double drums as the steel cable was being wound on; each hauled a drag door and one side of the trawl up to the side of the boat.

As soon as the doors broke water, we could see that the net was hanging down in a V-shape instead of the usual U-shape it assumed when things were normal.

Paul stood at the forward gallus, and I manned the after one as we waited for the doors to be "two-blocked" in their frames. As soon as that was done, we secured each door with a heavy hook on the end of a chain. Then Paul and I, almost in unison, hollered, "They're hooked! Drop the doors!" The men at the winch slowly released the brake wheels and, as they did, the chains took the full weight of the doors. We then unhooked each door from the tow wires, and Norm and Moe began to wind on the cable "legs" that led to the net. Usually the legs were open as they led to the top and bottom "sweep" of the trawl. This is the net's mouth, or front, which moves along the sea bottom. Now, however, the legs were wound tightly together. It became more and more obvious that we were going to have our hands full with whatever was hung up in the gear.

When the mouth of the net broke water, Paul yelled, "Look at the smoke!" Sure as could be, there was a wisp of smoke bubbling to the surface from the middle of the net, which was hanging straight down, about twenty feet below the surface. The smoke had to be a chemical reaction in whatever was fouled in the footrope or "chain line," which rimmed the very bottom section of the net, securing it along the bottom. *The only thing that would cause this phenomenon,* I thought, *would be military ordnance.*

I bellowed out, "Stop the winch!"

My crew had come to the same conclusion that I had, and Norm was already turning on the brakes and throwing the winch out of gear.

Since we had stopped the headway of the net, it began to settle straight down, right under the *Dorothy & Betty II. If I had my druthers,* I thought, *I'druther not be in this situation.*

No matter what sort of ordnance we had snagged, if it was "live," it could blow the boat and all of us aboard into little pieces. The problem with live ordnance at sea is that if you continue to bring it up to the surface, it can explode from the change in water pressure; if you release it, the increasing pressure as it sinks can also cause it to explode. There was also a proximity device on some of the later ordnance. These fuses would go off when they got near a target. You were damned if you did and damned if you didn't.

We were fishing in an area that had seen a great deal of naval warfare during World War II. German subs were seen, some were sunk, bombs and depth charges were scattered over the seabed all around Block Island. A number of convoy ships had been sunk during the war, and several planes had crashed while involved in finding and sinking the submarines; these, too, went down with weapons aboard.

What was hanging below our boat might be just a practice weapon, as well. Scores of these had been dragged up by just about every fishing vessel on the eastern coast. We were alone offshore, with no other dragger in sight, and there weren't any other boats that could be counted on to rescue us if we developed a serious problem. That did not make my decision any easier. Before taking any further steps, I called the Coast Guard at Point Judith, our home port, to advise them of our situation. At the very least, I wanted someone to know what might have happened—just in case I made a mistake in my choices.

The radioman at the Point said, "Hang on, Skipper; the chief wants to talk to you."

"Hey, Cap, what have you got yourself into?" he asked, just as you would expect from an old, experienced veteran.

"I'm not sure yet," I answered, "but I'm going to bring it aboard, or I should say, try to bring it aboard. I've given our bearings to your radioman, so if a problem occurs you'll know where to look," I responded. "As soon as we can see what we have, I'll call back with a description so you can log it in to see if it is live ordnance."

"Okay, Cap, we'll be standing by, and I'll keep this channel clear of any traffic until we hear from you. Good luck," the Coast Guardsman said in a slightly more serious tone than in his first comment.

"Let's go, fellers," I told my crew. "Haul her up slowly. I don't want to slam whatever it is into the hull."

While I was talking on the radio to Point Judith, I had been slowly turning the boat around so that we would be lying downwind of the net.

This would keep the gear from fouling the propeller. I then stepped out of the wheelhouse and stood by the after gallus frame to watch as the trawl was slowly raised up to the starboard side of the boat. Before the mouth of the net broke water, I could see what our problem was, and it did nothing to ease my concerns or my fears.

"It's a torpedo! Stop the winch! Let's see if we can get a line around the end of it to keep it from slipping down the net!"

Although the ordnance appeared to be thoroughly entangled in the twine, I was afraid of its breaking loose and ripping the entire net in two, possibly tearing itself free. It then could fall toward the ocean bottom and would very likely explode from the pressure change or upon striking the sea floor. Paul was quick to grab a nylon hawser, and he reached overboard as far as he could. As fast as I had seen anyone move, he had two half hitches around the torpedo's protruding propeller and steering mechanism.

"Okay, bring her up slowly," I said. "Paul, hand me the rope, and I'll take a strain on the torpedo. All right. She can't fall free, so let's try to get the net aboard."

We hauled the wings of the net to the gallus frames, then Paul and I each dragged the quarter ropes up to the open, fairlead blocks hanging from a cable rigged taut between the frames. We then passed the bitter ends to the two men on the winch. They, in turn, whipped the quarter ropes onto the winch heads and began to slowly haul the mouth of the trawl up to the side of the boat. As the net inched upward, the head of the torpedo began to swing forward and rise to the surface as the twisted twine encased it and brought it to the rail.

"Easy, fellers," I said. "Easy, dammit! We don't want that sucker to hit the hull! Let's just stop for a minute and think this thing through!

"Paul, you're the most agile. Do you think that you could climb out onto that mess of twine and slip a strap around the nose of that thing so we can hook the falls into it? That will keep it from slipping, as well as preventing it from striking the hull."

"Yeah," answered Paul, "but what if that thing blows? I'll be right on top of it!"

"Paul, think about it. We're only a few feet from it ourselves; if it goes off, we won't even know about it."

"I guess so," Paul said reluctantly, and overboard he went. He was hanging onto the twine that had been hauled out of the water, and when the boat rolled in the ground swell making up, he came near to getting dunked.

Quick as a wink, though, he had the strap around the head of the smoking beast, and Moe handed him the block with a large hook attached. It was inserted into the two ends of the nylon strap. I took a strain on the block and falls, then slipped the line around the winch head. As the strain was slowly increased, the torpedo rose higher and higher, away from the side of the boat.

"Paul, come back aboard," I said. "I want to get this thing on deck as soon as possible. The wind is freshening and if that little darlin' gets to swinging too much, things could get dicey." It didn't take Paul long to comply, the color coming back into his face as soon as his boots hit the deck.

We had to slack off on the line holding the stern end of the torpedo as we hauled the business end of the thing into the air. As we did, the ordnance began to lie across the deck, so I began to lower the falls.

"Norm, tie another line just below the strap that Paul put on the torpedo, run it out the scupper, then tie it down," I said. "I don't want that thing to begin slipping overboard as I let the falls down."

The *Dorothy & Betty II* was a little more than seventeen feet wide, and as our unwelcome guest was lowered and secured, its head hung out over the port side, and the propeller was inboard just enough to allow us to work on it.

"Norm, as soon as you can, try to find some identifying numbers off that thing," I said. "Write down any names or words or anything that will let the Coast Guard track it down. We need to see what kind of problem we have on our hands."

Just as soon as Norm retrieved the needed information, I called the station at Point Judith. The chief sounded relieved even as he joked about my still being in one piece.

"Hey, buddy, you just get the information we need."

"We're heading in now, Chief. With a fair wind we should be in the Point in three hours. How about getting a demolition crew from Quonset to take this thing off our hands when we get in? I don't want to lose any more time than necessary; I'd like to get back out as soon as we can."

"Will do, Cap, and I'll get back to you as soon as Washington tells me what you have on deck."

"Okay. Thanks for your help," I responded.

"Let's haul the wings and the doors aboard," I called out to the men on deck. "I want to get under way as soon as possible."

This job was done quickly, and as soon as we were heading for home, Norm took the lead in unshackling the end of the wings from the top and bottom legs. He wound them onto the winch drums to get them out of the way. Now he could begin the tedious work of getting the gear free from the torpedo. He started with the front end of the weapon and gradually worked his way down the length of it. This was no small task, as much of the twine was twisted as tightly as a cable, and he did not want to just slash it, as that would mean hours of work mending later on. So the three men worked out a system of rolling the twine and then sliding the loosened mesh toward the propeller mechanism that lay just inside the bulwarks.

By this time we had steamed the length of Block Island and could see Point Judith about fifteen miles to the north'ard. The radio crackled alive, and the very worried voice of the chief came on.

"Hey Cap, how are things going?"

As he was speaking, I was hanging my head out of the wheelhouse window to see what progress was being made by the men on deck. Moe was holding two of the propeller blades, and he would turn them slowly to the right as Paul and Norm worked the twine off. It was cut the twine and turn the propeller, cut and turn. "We're making good headway, Chief. We've nearly got the twine out of the propeller. We should have it clear by the time we get to the breakwater. What did you find out?"

"Cap, that is live ordnance you have on deck; I repeat *live* ordnance. And, Cap, that thing fires after so many revolutions of the screw. Did you copy that, cap? Don't turn the propeller *at all!*"

As those words were leaving the chief's mouth, Moe moved the screw one full turn to the right.

I screamed out the window, "Don't turn the prop! That's how it fires!"

Moe looked up and started to turn another revolution while asking, "What?"

"Moe, do not turn that propeller! *Stop!*"

The seriousness of the news given to us by the Coast Guard finally sank in for each of us, and you could see a white pallor rise into each face.

How many more turns of that screw would it have taken to blow us into oblivion? Thankfully, we would never know.

"Moe, let's just stop everything for a while," I said. "How about making us a pot of coffee? I need something to calm my nerves. If I had a jug, that would do nicely, but your coffee will have to suffice."

"Well said, Cap," said Norman. "When we get into the dock, I'll just slash the rest of the twine away from the screw and take time later to mend the net. I don't want to do any more than just look at that sucker until we get into the Point."

Captain Sam Cottle took the helm of his Maine-built dragger, the Dorothy & Betty II, *at the age of twenty-two, and fished her successfully for some fifteen years.*

Paul quietly concurred by going below with Moe.

As we steamed closer to Point Judith, or rather, the harbor at Galilee, I checked periodically with the Coast Guard. I informed them of our progress and inquired as to any precautions being taken to protect the hoards of sightseeing tourists, as well as the score of fishing boats and yachts in the harbor.

"Well, Cap, I've spoken to the munitions people in Quonset Point," said the chief. "It's their opinion that what we do or don't do won't matter a damn. They don't know why that torpedo hasn't sent you heavenward before this. So, based on their expert opinion, I don't plan to do a thing. If any announcements are made, we will have a panic on the beach, and if you notify the 'white boat' people, there will be utter chaos in the harbor.

At any rate, cap, if anything untoward does happen, we won't be around to blame, now will we?"

"That sounds reasonable to me, Chief," I answered. "By the way, I think we would have greater success in getting that torpedo off our deck if we could lie alongside the State Pier. Do you think you could pressure a few boats into moving enough to allow us room to squeeze in? We're just outside the West Gap and will be turning into the channel in about five minutes. We'll see you dockside."

It just amazes me that among the thousands of sightseers out on the breakwaters on both sides of the channel—on the beaches and along the bulkhead—not one person noticed this fishing boat listing to port with a torpedo sticking over the rail, the body of it lying across our deck as we slowly entered the breakwater and edged up along the finger piers, arriving at the State Pier in Galilee.

We had finally made it back home. The danger, we thought, was behind us, and the rest would be a piece of cake. You know that old saying, "Ignorance is bliss"? It really has some merit. Our real adventure was about to begin.

Once we were abreast of the State Pier, it was obvious that every inch of the dock was embraced by every conceivable type of vessel ever launched. Then I saw something that would surely solve our nerve-wracking problem: Taking up about one-third of the north side of the pier was a large Army Corps of Engineers barge that was being used to dredge the channel around the State Pier and the finger piers. It had a massive boom, fully rigged with cable, blocks, and hooks.

I blew a couple of short blasts on our air horn as we eased alongside the barge. One old—very old—crusty-looking deckhand stuck his head out of the galley. This was located aft of the wheelhouse and boom-control station. He had a cup in one hand, a well-chewed cigar in the other, and a wooden match sticking out of his mouth. I wasn't sure in what order he planned to use any of those items, but I really didn't care.

"Are you the captain of this vessel?" I asked with tongue in cheek. "If so, I have a favor to ask of you."

"Ask away, son, but I ain't no captain. Last captain I knowed was old

Black Jack Pershing that time we crossed the Mexican border. But what can I do for ya? I'm the only one aboard and will be for a few days; the crew was sent out on another job."

"Well, Cap," I said, "we need to get this torpedo lying across our deck up onto the pier and then onto a truck being sent from Quonset Point. If my crew gives a hand, do you think that you could do that for us?"

"I don't see why not," he answered. "I can handle this boom in my sleep. Why, I could put a quarter into your back pocket—if I had a quarter. You send your boys over and we'll take the dredge bucket off the end of the boom and hook up a simple block and falls."

I called the crew back to our wheelhouse and gave them a word of caution about the condition of this old salt.

"He just may have more than coffee in that mug," I said quietly.

Then, addressing the old crewman aboard the Army Corps barge, I queried, "Now, we're going to put a new nylon strap around the point of balance on the torpedo and we will have steadying lines on both ends to keep it from twisting and possibly hitting anything between here and the top side of the pier. How does that sound to you, Cap?"

"That sounds great, but just call me Carl; I get confused when you call me cap. You ready for me to lift her yet?"

"God, no. I want to wait for the navy truck to arrive. I don't want to handle this ordnance more than once."

"Now, Cap, did I just hear you say that this was ordnance? Like live ordnance? Like one that can explode?"

"Yes, sir, that's what the Coast Guard told us when we called in the numbers we found on it. Why?"

"No reason, but I have to get me another coffee. I'll be right back." And back he was in about five minutes. This time, however, he had a whole thermos of "coffee."

The more he drank, the steadier he seemed to get, but this change in personality began to concern me. We needed a reliable hand in removing this weapon from our deck, putting it safely on a truck, and—we hoped—getting it out of Galilee.

A half-hour passed before the huge flatbed sent by the navy arrived. The vehicle carried lots of blocking and ropes, but only two seamen, both of whom looked as though they had just finished boot camp. The driver leaned out of the window of the truck and, with a distinct quiver in his voice, asked, "Is this the place we're to pick up some kind of bomb or something?"

Norman said, "Yes—or something."

I hollered up to the driver that he was to parallel-park his truck so that the flatbed would be in a direct line with the ascent and descent of the torpedo as we moved it up off the deck and over the barge and the pilings on the pier, then delicately laid it down on the bed of the navy truck.

"Torpedo? No one said anything about no torpedo, did they, Charlie?" he asked the other baby-faced swab. His friend answered, "No, no, they didn't."

"Well, you're here, and you're going to take this sucker out of our lives just as quickly as we can give it to you," I said with more than a tone of anger.

"Carl, have you got that engine fired up enough that we can use that boom?"

"Yes sir! Let's do it," he slurred as he took a long pull on that coffee cup.

Norman and Paul were holding lines on either end of the torpedo. I climbed on top of it and inserted the large hook at the end of the cable coming down from the end of the boom.

"Okay, Carl, lift it as slowly as you possibly can, and no quick motions! Understand?"

One more quick gulp from the coffee mug, and Carl reached down to pull up on one of the myriad levers in the control booth. I was still straddling the torpedo and was in the process of lifting one leg over it when the massive and very deadly weapon shot up in the air, throwing me into a tangled mess of twine and rope. Norman and Paul had both lines pulled out of their hands, and they were yelling to me and then to Carl.

The torpedo was suddenly lifted into the air about twenty feet above our heads and, as it went up, it began to twist violently. I yelled at Carl to bring it back down within our reach. In response, he must have released the winch brake, because the torpedo started descending at a very rapid rate. Just as I thought it would crash into the *Dorothy & Betty II*, Carl yanked on the brake lever, and the weapon stopped inches above the deck. As Norm and Paul regained control of the guidelines on each end of the device, Carl tipped his thermos right up into the air and gulped its contents as a quantity of elixir ran down his chin.

"I've got it now, Cap; I'm in full control. Let's go one more time." And as he said this, Carl slowly raised the boom and its volatile cargo heavenward, and he also began to turn the base of the boom mechanism. We watched in horror as the torpedo began to swing into the pilings on the State Pier, which were spaced about ten feet apart. The torpedo was twenty feet long.

Wham! went the ordnance as it struck the pilings, bouncing back and striking them again. Carl then deftly raised the cargo to clear the pilings, and he continued the turn of the boom.

Now the torpedo at the end of Carl's cable struck and smashed the side of the truck. He raised the boom a little higher, turned it a foot or so inboard, then released his brake control completely. As the two swabbies jumped from the truck, the full weight and potential destructive powers of the torpedo landed on the bed, almost as planned.

"Okay, you guys, get that damn thing out of here now!" I shouted. The blue jackets jumped into the truck, put it in gear, and roared down the "escape road" out of Galilee toward someplace called home.

Norm, Paul, Moe, and I looked down at Carl. He was old when we first met him, but he had aged many more years as a result of his generous offer to help us out of our awkward situation. We all approached him with joyous acclaim and commended him on the very brave thing he had done. I could think of no more meaningful expression of thanks than to invite Carl and my boys to the Bon View for a stiff drink in appreciation for our extension on life.

"Just think," I said. "We may live to do something like this again. Another round, barkeep!"

THE LARGEST HAUL

We were steaming north up the West Passage between Bonnet Point, Rhode Island, and Conanicut Island—commonly called Jamestown, after the only town on it. This course would take us into the larger body of water known as Narragansett Bay. Our plan was to arrive on the west side of Prudence Island shortly after daylight and wait for our plane to arrive out of Theodore Francis Green Airport, near Providence.

This was a whole new venture for me and my crew. We were searching for pogies, the colloquial name for menhaden. These fish contain great quantities of high-grade fish oil and usually brought a good price per ton back in the early 1960s.

A number of boats out of Point Judith had been successful in following

Before he could switch from bottomfish to pogies (menhaden), Captain Sam had to find a seine boat to handle the huge net needed to encircle big schools of this oil-laden species. He bought this never-named forty-three-footer in York, Maine. Double-ended, she was set up to steer in only one direction.

the large schools of fish needed to meet the financial outlay of a seine, a seine boat, an airplane with a pilot and fish spotter, radios for communication between the boat and plane, and three or four times the number of crewmen usually found on board for dragging. Reluctantly, I have to acknowledge that the huge shares being earned by the pogie boats were the primary motivation for my own expenditure in rigging out the *Dorothy & Betty II* for this new fishery.

Once I had made the decision, I had to spend weeks chasing up and down the New England coast looking, first of all, for a seine boat. It had to be a double-ender, that is, having a hull that was pointed fore and aft. And it had to have the steering mechanism that is peculiar to seine boats: it only turns to starboard. This was completely different from anything that I had encountered before, and it took some getting used to. I had to develop a mental image of just how large a circle I needed to make to surround a school of fish without cutting it in half, as well as how to hold a course that would return me to the mother ship.

Up in York, Maine, I found a forty-three-footer that fit the bill. It had been a few years since any thoughtful soul had put a coat of paint on this old girl or had hauled her out to repair any damage incurred while she lay against the dock and chafed on the rocks. That was our first task after bringing her to Point Judith. As I write this, it dawns on me that I didn't give this worthy old girl a fitting name. I had never before failed to do that. I have even named the skiffs and other small powerboats that I've owned over the years.

After the needed hull repairs had been made, we put a bright coat of white paint on her hull and contrasting "red lead" on her bottom. Then came a new shaft and propeller, repair of the reverse gear, and a tune-up of the engine. All did wonders for her personality. Only one task remained: putting all-new stainless-steel rigging on her mast. When we brought the boat home from the shipyard, she sparkled in the sunshine; she was probably proud of her new looks.

While all this was going on, the biggest and most costly part of the conversion was taking place down at the dock in Galilee. A tractor-trailer pulled up to offload a single length of fine-mesh twine. It was by far the biggest pile of netting that I had ever seen in my life. If you could have run it out in a straight line it would have covered nearly a mile. Of course, that would never happen because the boat onto which the twine was going to be loaded would only turn to starboard. But it would make one very big circle.

Plastic floats had to be secured to the top line, and lead weights to the bottom one. Large metal rings used to close up the net when full of fish were also hung along the bottom. Through these rings ran a strong nylon rope secured to a seven-hundred-fifty-pound lead weight. When this was released, the bottom of the seine quickly closed, capturing the fish within the circle. If you envision a lady's purse upside down, then you get the picture —hence the name "purse seine."

One last piece of equipment was needed to haul this massive body of wet, fish-laden twine. And that was a large hydraulic "powerblock" that hung off the boom on the mast of the seine boat. Once the net was set around a body of pogies and "pursed," the twine was fed into this powerblock, which would grip the twine and wind it back into the boat. The seine boat crew would separate the headrope from the bottom line as the twine came off the block, getting the net ready to be set out once more. As this was taking place, the primary powerblock would "harden" the body of fish, forcing them into a smaller area from which they could be bailed onto the mother ship with a big, hydraulically operated dip net or by a huge vacuum pump that would suck them directly into the fish hold.

Years ago, when southern pogie boats from Virginia and the Carolinas worked their way up our coast, each vessel was rigged with a very high mast topped by a crow's nest. A number of men would spend hours looking for schools of menhaden and then directing the seine boat toward the fish.

As previously noted, we planned to use a plane that carried a spotter in addition to the pilot. The spotter had a working knowledge of the boats involved and the likely movements of the fish being encircled. The plane could cover miles of water where the fish might show up, and this would save the mother ship many hours of steaming.

Finally, with all the specialized equipment in place, the crew trained, and the pilot and spotter coordinated with the mother ship and its captain, the *Dorothy & Betty II* was ready and raring to go.

We found and set on a number of schools of pogies over the first three weeks of fishing, with an impressive degree of success. Obviously we had a few foul sets and our timing needed to be honed, but all in all, the crew became very proficient in a very short time. The volume of fish increased each time we went out, and we felt ready for the big test. We hoped it would be soon.

Then came the day where this tale began, when we were steaming up the West Passage heading for Narragansett Bay. The sun was about to break over the horizon, and by every indication, the weather would lend itself to a banner day for fishing.

The crew was in the fo'c'sle having breakfast. I was at the wheel as we steamed up toward the Jamestown Bridge. The water was smooth as glass, with a shiny, greenish cast as the new rays of sun flicked across the water. I had just tried to raise Paul Champlin in the plane to confirm the coordinates of the boat's position but I'd had no success. He probably hadn't yet taken off, so his radio couldn't receive me.

Just as I turned back to the windows of the wheelhouse, the surface of the channel exploded with breaking fish. The water was a frenzy of white water and silvery bodies as far as the eye could see. I would guess that the school of pogies before me was as large as a football field, if not bigger.

I blew the air horn and stopped our headway in the same motion. As the men poured out of the companionway, I just pointed over the bow. As if we encountered a phenomenon such as this every day, the crew immediately took their places and began the smooth transition to an operation of efficiency not often seen on a fishing boat in these waters. I slowly steamed closer to the massive body of pogies. The glare of the early sun blazing across the water would have kept us from seeing these fish had it not been for their violent frenzy and the resulting white water. This was an excellent demonstration of the importance of a spotter plane under difficult conditions. Just as these thoughts were running through my mind, the radio crackled into life. It was Paul saying that he and the spotter had just taken off and would be meeting us as scheduled. I told him to land the plane and recounted what was going on. I suggested that he go home until he heard from us later on.

Bringing the *Dorothy & Betty II* as close to the pogies as I could without spooking them, I shut down the engine after checking the depth of water, the speed of the tide, and the direction of the wind. All of these would have an effect on the drift of the mother ship as all hands, including myself, went into the seine boat or one of the other work skiffs.

At the helm of the seine boat, I began to steam toward the huge body of fish. As we arrived at its fringe, I called out to a man in the bow of one of the work skiffs, telling him to slip the towline. As they fell back, they used their outboard motor to turn and head away from us so that they in effect became a brake as we in the seine boat began to circle the fish. The tension supplied by the skiff hauled the end of the big net over our stern at a rapid pace—lead line and rings toward the inside of the circle, and floats toward the outside. It was as pretty a sight as a fisherman could ever imagine.

The second work skiff had been released at the same time as the first, just so that it was available in the event something went awry. With their high-powered outboard, they could quickly get to the trouble that needed attention.

As we made our circle, one man climbed to the masthead and gave directions for the gradual turn to starboard that would encompass as many fish as possible and not cut off part of the school. The body of twine was whistling overboard, and as its end neared, we were coming full circle, about to meet up with the first work skiff and join the ends of the seine. This was rapidly done, and just as that was completed, the man at masthead called out, "They're still on the surface."

I yelled, "Drop the lead!" The seven-hundred-fifty-pound weight was quickly released, and as it hit the water the seine boat rocked suddenly from side to side. As the lead headed toward the channel bottom, one end of the seine was lifted into the powerblock and the job of closing up the bottom of the twine began, as did the hauling process.

All hands now climbed to their assigned positions. Some were to clear fish, debris, and any twisted stretch of bottom line and top line from the gear on the water side of the block. Others worked the inside flow of twine as it came out of the block, and they were to haul the top line (with the floats) one way and the bottom line (with the lead) another. By the time we got to the hardened-up body of fish, all of the twine hauled was ready to be set again. As this was going on, the second skiff was cruising back and forth, waiting to address any problem that might develop.

Until the bottom of the seine was closed, we had all waited anxiously to see if the fish inside would dive to the bottom, find some opening, and all escape in minutes.

If this had happened, then the hard work and timely action on everyone's part would have been to no avail. The better part of a day would have been wasted, and all hands would have been exhausted and demoralized.

On this set, however, the rings along the bottom of the seine closed up quickly without any foul-up. In an instant, it became obvious to us that the fish we'd set on were in the net, as we had no more than hauled in about half of the twine when we could feel their full weight and the cork lines began to dip underwater. If this kept up, the pogies would quickly sense the way out and overwhelm the low spot in the twine, allowing most of them to stream over the corks.

It was time for me to use the radio. "Norm," I instructed, "bring your skiff up to the far side of the set, and lift the twine over your bow. Moe, take your skiff, run down the cork line about two hundred feet, and do the same thing. When you're both in place, let me know."

About ten minutes went by, then both skiffs acknowledged that they were in place, that the low section of twine was up out of water, and that the tide of escaping pogies had stopped.

Again I spoke into the mike: "As soon as we harden up enough fish to begin bailing, I want to slowly work this net inshore. These pogies will begin to suffocate and die, and in this depth they'll take the twine to the bottom and the boats along with them. I'll give you a holler when I want to begin. Meantime, stay where you are, and keep your eye out for any other problems."

Norm, Moe, and the men with them assured me that they were ready to move when I called for it. Their work skiffs were powered with fifty-horse outboards and were capable of moving that net full of fish, albeit slowly. As this was taking place, a couple of sportfishermen came close to see how we were doing.

I called one of them over and asked if he could take me to the *Dorothy & Betty II*. He quickly agreed and came in close enough to let me jump aboard.

"How would you have gotten back to your big boat if I hadn't come along?" he asked.

"I would simply have told one of my work skiffs to pick me up," I replied, "but I would prefer that he stay where he is."

We soon came alongside the dragger, and as I climbed on board I asked him how much he wanted for his troubles. "I'd like a couple of bluefish when you get the net in," said the angler. "I'd also like to stay aboard, if I could, to see how you do this!"

"Okay," I answered. "Secure your boat to the stern bitt, and I'll get under way. When we get to the set I'd like you to stay up on top of the doghouse. That'll keep you out of the way." Within a few minutes I had the *Dorothy & Betty II* alongside the seine boat as the crew continued to harden the twine. A couple of men came aboard the dragger to help me secure the two vessels together by hauling some twine up to the side and tying it down in several places.

Now the large body of fish lay between the mother ship and the seine boat, with the work skiffs on the far side of the haul. I called two men out of the seine boat to handle the dip net we used to scoop pogies into the fish hold. It was shaped a lot like the much-smaller scoop net that I'd used many years earlier in bailing fish out of my grandfather's fish traps. That one held about one-half bushel; this one handled about a ton.

The bale was a large round steel ring that had a deep pocket of twine hanging from it. The net bottom was fitted with a series of rings held closed by three slipknots that were released as the scoop was swung over the open fish hatch. The big dip net was directed by a long wooden handle in the hands of a very strong man who had to be able to swing it about as well as push it deep into the roiling mass of fish.

When the scoop was full, a signal was given to a man at the winch who handled a whip rope that ran from the rim of the dip net up to a block aloft, then down to the winch. (Later, the industry adopted a big vacuum pump linked to a huge hose that was lowered into the mass of fish, sucking them up and moving them into the mother ship.) It took real men to operate our system. It sure did.

We began to bail and bail and bail some more. *The Dorothy & Betty II* could accommodate seventy-five thousand pounds of fish belowdecks, but in no time at all the hold was full. At that point, we prepared to load up the decks. We secured two-by-ten-inch oak planks to the top of the bulwarks running along both sides of the deck. Checkers forward and aft kept fish contained within the designated area.

Soon the deck was full, too, right to the top of the "monkey rails." The *Dorothy & Betty II* couldn't take another bushel. I looked over the side and saw that we still had about a foot of freeboard. She loaded right level with the burden, and I was delighted with the way the boat handled all that weight. Now I would see how she managed under way with this huge payload.

The crew began to harden up the remaining body of fish, securing the twine aboard the two work skiffs and seine boat. When this was complete, I turned the *Dorothy & Betty II* around very slowly until I lay side by side with the seine boat, then I put her in gear and carefully edged the whole mass of fish and boats toward the Jamestown side of the channel. While Norm held the wheel, I watched the depth sounder and checked the chart until I felt we were in shallow-enough water that when the remaining fish died and settled, they wouldn't be able to sink the net—and the boats attached to it. The pogies could settle down to the bottom but the seine boat and work skiffs would still be able to hold the twine high enough to keep the fish contained.

Fortunately, we had sandy bottom beneath us and not mud that would bury the twine, or large rocks that would tear it. The only serious concern I had was the weather. The marine radio and the glass all indicated good weather, but it was common for a sou'westerly breeze to pick up on these summer afternoons. That would give rise to a heavy ground swell that could roll right up the channel, as there was nothing but open water to the south'ard.

I had several of the men come aboard the dragger to pick up some coolers full of ice and food and plenty of cold drinks, including water. They had several radios for communication with me and with the Coast Guard, if needed. I would be checking in with them at various stages of my run to New Bedford, the closest fish plant that was buying pogies. This seventy-mile round-trip would take upwards of ten hours because of our huge catch, and I had no idea how long I would have to wait to offload the fish.

When all was secure, I had life jackets passed out to each man—just in case. I then had a couple of men lay two wooden planks on top of the fish on deck, allowing me to move between the wheelhouse and the fo'c'sle doghouse, as in time I would need to go forward for coffee and a meal. The crew put cleats across them so they would not drift apart while I was walking, dropping me onto the deckload of pogies. I also had them run a lifeline from the winch in front of the wheelhouse to the forward mast to keep me steady as I made my way across the fish.

Bidding farewell to my entire crew, I slowly backed away from the seine boat so as not to cause a wake or upset the skiffs, and turned the bow of the *Dorothy & Betty II* down the channel. I was heading toward Beavertail Point, where I would take courses of 162 degrees to the Brenton Reef buoy, then 110 degrees to Buzzards Bay, and on into New Bedford.

I was very careful to keep away from the rail during the trip. I didn't want anyone to find a deeply loaded fishing boat sailing merrily along with no one aboard. Fortunately, pogies harden up considerably when they die, making my walkway of wooden planks fairly secure. Hake or whiting packed as deeply as this deckload would slosh back and forth—and overboard—with every roll of the boat.

After an uneventful trip to New Bedford and unloading the fish, I headed back to Jamestown. I arrived at 6:00 P.M.—just fourteen hours from the time the pogies first broke water. The crewmen were all lying on the twine, fast asleep, but they jumped up when I laid on the air horn. Momentarily afraid that they were about to be run down, the men were quickly relieved when they saw that I was the one edging up on them. I, in turn, was happy to see that they had survived a long day on a wet load of seine twine with a large body of fish still confined in the net, just below the surface.

We quickly adjusted the load of fish so we could begin bailing them aboard the *Dorothy & Betty II*. Now that all the pogies were dead, it took very little effort to bunch them together. Once more we filled the dragger's hold and laid on another deckload. Trouble was, we weren't finished. Because of weather conditions, tides, and possible boat traffic, I felt it best to leave all the men on the seine boat with the remaining fish. It looked as though we would have at least a hold full on the third trip.

With the *Dorothy & Betty II* loaded down much as she was on the first run to New Bedford, I was ready to begin the return trip. I had offloaded 105,000 pounds of pogies that time and had nearly that many now. I told the men to tighten up the remaining fish left in the seine, and then we worked them into somewhat shallower water. We made sure that the running lights were working on the seine boat, as darkness would soon envelop it. Concerned for the safety of the men in the event of heavy weather that would threaten a boat burdened with a big load of dead fish, I told Norm to dump the pogies if he ran into any kind of problem.

I then supplied the crew with food, water, and soft drinks, and filled their thermos jugs with fresh, hot coffee. By this time most of them had put on clean, dry clothes, as well. When all seemed secure, I swung off for the second trip. The time was 8:00 P.M.

Thankfully I had a dependable autopilot and an excellent radar working for me. This equipment relieved me of any undue concern about boat traffic and the weariness that comes from steering for too long. The trip was uneventful, but when I got to the processing plant in New Bedford, three other boats were already waiting to unload, and the plant crew said that they were having some problems with the pump. They predicted that it would be three or four hours before they could get to me. I told them I was going to turn in and asked them to call me when they were ready. This they did, just three hours later. By the time I was offloaded and under way on the return run, the time was around 4:00 A.M. The second load tipped the scale at ninety-eight thousand pounds. Two down, one to go.

I couldn't believe the weather. The wind had been light from the sou'west for two days, and we didn't have any fog. I was not that familiar with New Bedford Harbor even in bright daylight, and it would have been a bear if I'd had to navigate in "pea soup." Hundreds of boats sail in and out of this port every day, many of them out of Fairhaven, on the east side of the bay. In late summer, the Fairhaven side had (and still has) great numbers of 'white boats' (yachts), and you never had any idea where they would be as they wandered around in the fog. I felt blessed by the clear weather and moderate winds and only hoped that it would last a few more hours—at least until I could relieve the crew aboard the seine boat. I was anxious to get them home.

As I cleared Buzzards Bay and headed west in the *Dorothy & Betty II*, the wind freshened some, but I didn't think it was strong enough to trouble the crew because they were much closer to the shore than when we started this marathon trip. Still, I was concerned.

All went well as I swung up the West Passage and met up with the crew. Since it was 9:00 A.M. when we secured the boats together, I told all the men to come aboard the dragger. I had prepared a big breakfast for them during the last leg of the journey, and it would be their first hot meal in two days. They were happy to get it, and the reward was well deserved. It was all I could do to keep up with their appetite. Once refreshed by the food, we were all anxious to finish up and get home, but that would take one more trip.

As we took aboard the last load of pogies, I called Paul, the spotter pilot who had been at home all this time. I told him that when we were done bailing, I would tow the seine boat across the passage to Bonnet Shores. There was a small dock there that we could tie to until I came back to pick up the forty-three-footer and take it back to Point Judith.

The last load came aboard quickly; the seine was laid out in preparation for another set; and the skiffs were secured, ready to be towed. Then we headed across to the Bonnet. When we arrived, Paul was there with a second driver and the two vehicles needed to carry all the crew. Norm had offered to accompany me to New Bedford to give me a chance to sleep on the final trip to the processing plant.

We offloaded without problems and were soon on our way back to the Bonnet to retrieve the seine boat and skiffs. With these in tow we headed back to Point Judith and home. The wonderful weather had come to its end, and there was now a stiff sou'westerly wind and a rolling sea. Towing three boats astern in good weather is always a concern, and in heavy weather it is a real chore.

We had a good ten miles to go to get in through the East Gap, and the worst time would be when we reached the whistle buoy and had to turn westward, causing us to run in troughs between seas. The seine boat and skiffs were rolling deeply, and the forty-three-footer was beginning to ship more water than I liked to see. Still, with such a short distance to go, I could not see risking more trouble by turning around.

Things were getting really tight when I spotted one high, rolling wave making up on the horizon. It would surely swamp the boats we were towing, but I didn't have room enough to turn into the oncoming seas. The middle wall of the breakwater was just too close. My best shot was to increase the throttle to the rusty notch and try to reach the breakwater's lee before the combing sea hit us broadside, more than likely sinking the seine boat. As I looked out the port-side wheelhouse door, the rogue wave hit the southerly bend of the middle wall and was crashing along as it worked toward us. Green water was enveloping the gigantic rocks that had been placed there a century before to protect seamen from danger.

I edged the *Dorothy & Betty II* by the east end of the middle wall and, rather than turning as I normally would have, I held a straight line for the beach in an effort not to slow our headway for a moment, hoping that this maneuver would pull the seine boat into the protective embrace of the Harbor of Refuge in time. As I watched, the main body of water roared across the opening of the gap and engulfed the short, curving arm of the East Wall, which projected out into this wild combing ocean, breaking it up into smaller segments that hit the beach behind the riprap with a mere whisper of its once roaring and devouring power.

As soon as the seas had missed our tow and crossed the gap, I made as tight a turn to port as I could, hoping to keep our precious cargo off the stony beach in front of the old and now-defunct Coast Guard lifeboat building. I then quickly turned hard to starboard to straighten out our towline, giving us greater power as we hauled the three smaller boats rapidly away from the shore. We were successful in our efforts as the towing hawser took up the slack and yanked our unnamed girl to safety.

I soon backed off on the throttle to let things settle down and to allow myself a breath of air. It seemed as if I hadn't taken one in the previous ten minutes. I knew one thing: my adrenaline would be pumping for the rest of the day. The seine boat and the twine aboard her had done a masterful job this trip, and they had deserved more than ending up as a twine-covered shipwreck on a shore that had witnessed hundreds of such disasters over the nearly two centuries that the Point Jude lighthouse had stood watch.

When we got out of the rolling seas, we swung up close to the inside of the bend in the offshore wall. As we slowed down, Norm was able to shorten our towrope to give us better control as we crossed the West Gap and proceeded into the breachway leading to our dock and a well-deserved rest. The crew was there to help us secure the seine boat, the work skiffs, and that dear mother ship, the *Dorothy & Betty II*. Their first question was, "How much did we have in that set?"

"Well, boys," I replied, "we unloaded 81,000 pounds on our third trip, so that makes a grand total of 284,000. Not bad for one set. That's the largest haul ever for the *Dorothy & Betty II*, but I hope there'll be many more to come."

As things turned out, this catch would be the largest we ever landed. Before we were able to go out again, we got word that every pogie-processing plant along the Eastern Seaboard had closed down. Herring and anchovies had hit off Georgetown, Guyana, and the west coast of Peru. Those countries were catching the fish, processing them into high-grade oil, loading it into tankers, and delivering the processed oil to markets in Boston more cheaply than we could deliver the raw product.

So, our venture into the pogie fishery lasted less than thirty days. Nonetheless, it was a good month, one in which we caught over one million pounds of fish. I sold my equipment—at a tremendous loss—to those who had the finances to weather the hard times.

Still, I count that month as one of the great experiences of my life, a costly experiment but certainly an important chapter in my ongoing education as a fisherman.

Yes, without a doubt, that last set was my largest haul.

Leaning out over the end of the "stand" (pulpit) that projects some twenty feet forward of the bow, the harpooner readies himself to strike a swordfish that is swimming left to right just ahead of the boat. The fish is "horned out," meaning that its dorsal fin and the upper lobe of its tail are clearly visible above the surface of the water.

THE STRIKER

It has often been said that horse racing is the "sport of kings," and I suppose to the masses of people who engage in this pastime, whether as owners or observers, it just might seem to be. I have been an observer of horse racing and an active participant in swordfishing, and in my opinion, there is no comparison.

If a king presented himself to the captain of a swordfishing boat and went to sea in search of the ever-elusive *Xiphias gladius*, and if he partook in the unbelievable thrill of harpooning one or riding the waves in a dory as the struck fish was retrieved, horse racing would quickly lose out as the king's chosen sport.

I have spent many an hour of many a day at the masthead of some very fine swordfishing boats, riding swells from Block Island, Rhode Island, to the Peak of Brown's Bank, off Nova Scotia, and beyond. I passed those hours alongside the most experienced fishermen on the Eastern Seaboard, men who had seen many strange sights and encountered many unusual situations. Such experience generally conditioned this kind of man to be unruffled in the face of dangerous or thrilling events.

Yet just let a swordfish "horn out" (show its fins) on a glassy-calm ocean surface, and such stoic men will scream like banshees in the misty bogs of Ireland.

It doesn't seem to matter how many times in a day or a season they see that beautiful metallic-purple fish gliding just under the water's surface. The adrenaline floods their bodies, and they all scream, "There he is!" If they were to release that volume of noise in a confined area, I would fear for the eardrums of all listeners.

In the evening, after all hands had finished cleaning the fish caught and icing them down in the hold, we would sit around the table in the fo'c'sle reviewing the day's events. When conversation began, you usually had to lean toward the man speaking, as he could not speak above a hoarse whisper. Each member of the crew, in turn, would recount his excitement in the same tone. After a few attempts, we would quit talking —or trying to—and turn in for the night.

As we arose the next morning and were waiting for coffee and breakfast, with nearly renewed voices, each man would swear to the others that he would not raise his voice the next time he saw a fish. That oath would last only until the first fins broke the surface. "There he is!" the roar would come forth as each man in the mast spotted the "horns." It mattered not if the swordfish was finning or underwater; the reaction was the same.

I thought that seeing one of these great fish must represent the peak of excitement until I was given a new assignment. It was aboard the *Athena* out of Galilee. Ken Winter was skipper, and the two men who owned the boat were deckhands. I was the cook and hold man. This was not the best site on the coast, but until something else came along, it would have to do.

We had steamed to the east'ard, toward the area just south of No Man's Island, in hopes of finding some butterfish.

The sea was flat calm, there wasn't a cloud in the sky, and the sun was warming up the surface water as nice as could be. Just after breakfast, while I was cleaning the dishes and the fo'c'sle, we made (spotted) our first swordfish. Ken blew the horn, and I raced up to the wheelhouse because he didn't want either of the owners at the helm.

Ken raced forward, which on that boat was no easy task as the deck was like an obstacle course. He continued tripping over this and that, and by the time he ran out on the stand, he was as excited as a bride on her honeymoon. He yanked at the lanyard holding down the harpoon, but it refused to release. By that time, we were running down the swordfish and it was passing under the keel of the boat. Of course, the fish was not about to allow that, so it dove toward the ocean bottom and was never seen again. Ken continued to yank and tear at the lanyard, keeping up a stream of cusswords that his priest had never heard.

Just as he cleared the harpoon, one of the owners yelled that another fish had horned out a few hundred yards south of our last sighting. Ken told me to turn in that direction as he picked up the pole and checked to see that everything was in order. He turned in the stand so that the pole was on his right side and raised it as smoothly as an old whaler. I brought the boat onto the fish with the sun behind us and it couldn't have been placed in a better position. Ken made a masterful throw that went deep. In fact, it went right on down the side of the fish into the ocean depths until yanked to a halt by the bib line tied to the near end of the pole. Another cloud of blue flames flew from the mouth of one of the best cursers in Point Judith.

This fish must have had a death wish, as it turned around and finned a hundred yards directly in front of the stand. Again Captain Ahab struck at Moby. Again the deadly shaft struck only water.

In a monstrous rage, Ken came aft, telling me to get below and make some coffee. Before it was perked, another swordfish horned out to his back. Ken hollered to one of the owners to go out on the stand. He did, and he missed—another blown shot. It was unbelievable, but now fish were horning out in nearly every direction. Ken swung the boat onto one after another. Both owners took turns as striker, and both missed by a country mile every time. I was so discouraged that I never left the fo'c'sle. I decided that if we weren't going to set the net and do something that I knew Ken could shine at—catching butterfish—then I would stay below and start dinner. If we did settle down to serious fishing, I might be too busy to prepare a good feed, so there was no time like the present. I had just finished putting the noon meal in the oven and was washing the breakfast dishes when Ken stuck his head down the companionway.

"Come on up here. We've got a fish horned out dead ahead, and I want you to take a shot at striking. You can't do any worse than the rest of us."

I dried my hands, lit a cigarette and stuck it in my mouth, and slowly climbed the fo'c'sle stairs. I stretched, looked around (but not at the swordfish), and climbed over the raised forward deck. Along the way, I checked the kegs, the warp, and every leather becket (strap) that secured the warp from the lily (dart) to the pole and along the rail aft, but I avoided looking at the swordfish we were now rapidly approaching. I sauntered out on the stand and into the semicircular hoop used by the striker, untied the pole, lifted it clear, and turned to the right until I was facing the direction of the fish. Then I lowered my left hand slightly and raised my right hand and the pole. Then, for the first time, I looked at the fish.

I still had the cigarette in my mouth, and the increasing smoke from my heavy breathing was burning my eyes, as well as hindering my view of the fish, which was settling slowly beneath the surface as we approached it at eight knots. I tried to spit out the cigarette, but it stuck to my upper lip and then turned around and seared my lower lip with its glowing red tip. I did not have time to put down the pole and remove the butt from my mouth with my fingers, so I yelled from the pain, lined the fish up with the pole, and struck.

Life became intense in the next few seconds. The harpoon left my hands, which flew to my mouth and tore the stuck cigarette from my lips, burning two of my fingers. The swordfish dove straight toward the ocean bottom, the pole settled quickly on the water's surface, and I was certain that I had missed, as had all the others. Then I saw and heard the warp sizzling out of the basket and watched in delight as the keg was yanked off the deck to splash into the water below. At that moment I knew for certain that the fish was hit. I had struck my first swordfish! The excitement was different from any other I had ever experienced, then or thereafter.

That fish, my fish, was tended from the mother ship. It was a risky thing to do because it was easier to pull out the dart, but Ken said that the dory wasn't fit to be lowered into the water, let alone to accommodate a man who would then haul the fish up to its weathered sides. One swipe of the sword against the length of the dory and it would pop open like an eggshell.

As quickly as that group of swordfish surfaced, they disappeared. We then settled in for the work that we had initially set out to do—drag for butterfish. We did it with reasonable success, and before dark we had swung off for home with a goodly haul.

Of course, I was still enthralled with the good fortune of having struck my first swordfish. It was a thrill I can and do relive upon occasion, and, as things turned out, it allowed me to enjoy special status as a striker in the years to follow.

As I relate elsewhere, I had the pleasure to serve with Jerry Adams—a striker extraordinaire. I have seen him lean into the stand, most of his body extended over the water, just to get an extra foot closer to a fish that was beginning to sound. He would put the iron down by the fish's spine, and the dart would toggle (turn sideways) on the bottom side of its belly.

I recall one time, we were steaming into heavy seas that dunked Jerry nearly to his chest in ice-cold water as we turned the boat onto a wild fish, which kept jumping out ahead of us and swimming in tight circles as we tried to approach. Despite the conditions, we got him royally.

On another trip we were near the peak of Brown's Ledge, off Nova Scotia, late in the season. The only fish we saw all day finned between us and the old 46, a huge schooner out of Lunenburg Harbor. I was steering the *North Wind* from aloft, using a loop of rope that ran through a series of fairleads and around the ship's wheel, down in the pilothouse. I had the responsibility of bringing our boat into proper alignment with the swordfish so that the striker could iron it.

I was in a good position to view old 46. I say old because she had seen eons of time at sea as a dory trawler, fishing the Grand Banks. She must have measured one hundred and twenty feet or more, and could have carried our boat on her forward decks and still have ample room for her fishing gear without being crowded. As she steamed along toward the fish on a converging course, I watched the two men in the wheelhouse. The doors on both sides had been opened and latched back.

One man stood on the port side of the wheel and one man on the starboard side. The wheel itself was so huge that a trough had been cut into the floor of the pilothouse, allowing the wheel to pass beneath the deck as it turned. The members of the crew were masterful in handling the vessel as they came about onto a fish. The men aloft, seven at least, had a horn system for giving directional signals. When they saw a swordfish they would sound the horn: one blast for a turn to port, two for a turn to starboard, and one for steady as she goes. As the schooner turned toward the fish, the crew aloft would blow the horn just one more time—when the 46 was right on top of the fish. I never heard more than that.

The men at the helm knew exactly how that old vessel would react—how long it took to turn her and how long it took to stop her from turning. Then the schooner was dead on the fish. The stand was the longest that I had ever seen. The striker was so far from the fo'c'sle that he had to take a lunch with him when he took his position in the morning. The real advantage he had was that he was out over the fish long before it could hear or see their boat or feel the vibrations of its propeller. As a result, the striker could just "punch" the fish instead of throwing the pole. That assured him of success almost every time the *46* came over a fish.

Our boat was shorter by sixty-four feet, was much lighter, and rolled and tossed with every sea. Using steering from the mast allowed the *North Wind* to yaw (wander) much more because of the slack in the long rope. It was much more difficult to feel the boat as she came out of a turn and we tried to straighten the course. So Jerry didn't know for certain if he would be throwing to his left or to his right. All of that took its toll on him as we put him onto one fish and then another through the day. But despite that distinct disadvantage, he was remarkable in compensating and being deadly accurate as he threw the harpoon.

As both vessels steamed full speed toward that lone swordfish, we came closer and closer together. The men on both boats stared first at the pair of fins and then at each other. Nearing the fish, the strikers on both boats stood up as straight as the iron at the end of their harpoons. Each man checked and rechecked his pole, looking as closely as he could from a distance at the lines leading from the dart to the warp basket and the keg. All was well, but each checked again. He hefted the harpoon, looked the length of it, then looked again as the fish horned out just ahead.

The old *46* was steadier on her course because of her length and weight, but we had a slight advantage because of our greater speed. It would all boil down to a test of nerves. Who would turn first—or, as the kids say, "chicken out"? I will confess that I was getting very nervous. I could clearly picture the *North Wind* being smashed into splinters when the *46* rammed us as we neared that choice prize—a single, very large swordfish swimming quietly in the North Atlantic.

"How much longer do you want me to hold this course?" I shouted to Jerry with a tremulous voice as I came nearly eyeball to eyeball with the men aloft on the old *46*.

"If you turn off one point, you're fired! Do you hear me?" Jerry replied.

One point ain't much when you're running before a following sea and the boat is yawing one way and then another. I was hauling ferociously on the steering rope to hold the *North Wind* straight. No, sir, one point ain't much. "I'll try, Cap," I yelled to Jerry. "I'll try."

"You will more than try; you will d*o!*" came his reply.

We were close enough that a man could jump from one vessel to the other with little effort, but still we raced onward.

With just a few hundred yards to go, Jerry picked up the harpoon and turned slightly to the right, to align the dart with the oncoming fish. When was no room to do anything but ram into the *46*, I suddenly heard one blast from their horn. The schooner turned to port just as Jerry raised the pole and drove it solidly into the wide back of that chocolate-brown swordfish, which jumped as the iron entered and exited, the dart toggling beneath its belly. It was hit hard, real hard. I turned quickly to starboard—toward the sounding fish—in order to kick the propeller away from the warp that was whistling out of the basket on deck and trailing astern. If the boat wasn't turned away from the dry, floating warp, it could easily get into the ship's propeller and we would lose the ironed fish. Once we had completed the turn, the keg went over the rail and my brother-in-law, Roger, jumped into the dory and prepared to tend the fish.

Jerry turned slowly around, lifted his face with a wide grin, and looked at me. "Well, I guess you keep your job. Good steering, Sam. Let's see what we have at the end of that warp."

It was getting late in the afternoon and we would be shutting down shortly, so I called down to Artie Main to take the helm from the wheelhouse. After our little contest I had an urgent need to "pump bilges," so I rapidly covered the distance from the masthead to the leeward side of the wheelhouse. It was going to be a long time before I would forget the "showdown on the peak of Brown's."

Although I did not get the opportunity to hone my skills as a striker while on the *North Wind* with Jerry Adams, I did appreciate watching him over the summer months. His control of his emotions and breathing, and his slow movements prior to the moment of releasing the harpoon, were something I wanted to simulate when I was again allowed to step onto the striker's stand.

When Roger hauled the swordfish Jerry had ironed on that fateful "showdown" day and we brought it aboard the *North Wind,* we all knew that we had just taken the biggest swordfish of the season. It was a real keeper, but I still questioned the wisdom of the contest. A short time later, however, the skipper of the *46* called Jerry on the radio to congratulate him on getting the fish. He said that he would guess it would weigh out at eight hundred pounds or more. He could tell that from seeing it underwater and while he was steaming at eight knots or faster. Those "Novi" boys really knew their business when it came to broadbills. (When we later unloaded our trip at the Co-op in Galilee, the dressed fish weighed out at eight hundred seventy-one pounds.)

We always ended our season off Nova Scotia, in late September. If there were any hurricanes in the area they always ended up on that part of the fishing grounds. If we were lucky enough not to get a hurricane, October normally brought on one nor'easter after another, and with the swordfish on the move, it wasn't worth the time and expense to fit out a harpoon boat for such a long trip away from home. So we wished our friends on the old *46* Godspeed and good fishing, and swung the *North Wind* to the south'ard, toward Point Judith.

A MIXED BAG

The showdown on the peak of Brown's marked my last trip with Jerry Adams, as I had been working to buy my own dragger. With any success, I would be walking the deck of my own rig and, come swordfish season, would be putting on a stand, one that would hold me firm as striker. It was nearly three more years before this came to pass, but I was finally able to buy the *Roberta Dee*, followed within a year by the *Dorothy & Betty II*.

Swordfishing is much like a fever, or even an uncontrollable habit. I know of crews that will literally throw all caution to the winds when someone sees a fish horned out off Block Island while dragging. They will quickly rig the boat with a stand and topmast and dories and kegs and warps, and away they will go. They will chase that dream all over the ocean just to hear one of the men yell out, "There he is!"

On the other hand, as much as I enjoyed chasing broadbills, I always remembered that bottom-fishing was going to pay the bank note on the boat each month and put food on the table at home. I was just too conservative when it came to choosing between what I truly wanted to do and what had to be done. So each summer, although I put the stand on, brought the kegs and warps aboard, and rigged up the steering rope from the wheelhouse to the masthead, I also brought aboard the drag nets and doors. No matter how glassy calm the water was on those summer mornings just before daylight, I always set the net overboard and began searching for marketable bottomfish, oftentimes to the grumbling of my crew. When the ocean was flat and the sun was shining high, the men would have gone aloft in a minute if I'd allowed them to.

We enjoyed much success in combining dragging and swordfishing throughout the summer, and I was not about to change tactics, no matter how much I felt the urge to spend all my time looking for the telltale fins of swordfish on the surface.

Perhaps the results of one good day will explain why my method was more successful than that of many others fishing the same area. We had arrived just south of Muskeget Channel, which runs between Martha's

Vineyard and Nantucket. This channel has strong currents that stir up the baitfish, becoming a nutritious soup that washes out into deeper water with the ebb and flow of the tides.

We were in thirty fathoms when we set the net on the edge of the channel, hoping that butterfish and, just perhaps, some swordfish would be feeding in that stream of bait. It would not be daylight for an hour or more, and the butterfish would be just a few feet off the bottom. They would continue holding that depth until the sun rose higher in the sky; by 9:00 A.M. they would be completely off the bottom and beyond the reach of our net.

I soon called the crew from their bunks with a short blast of the air horn, and as the men came quickly out of the companionway, they made their way to their assigned stations. Moe released the hookup that had held the tow wires close to the stern; as he did so, Norm and Paul began hauling in the wires with the winch. Moe then went forward to hook up

Crew members tend the winches aboard the author's eastern-rigged (house-aft) dragger Dorothy & Betty II *as the net is hauled back.* .

The last step in the operation is "two-blocking" (snugging up) each heavy, wooden trawl door in its steel gallus frame.

the forward door when it broke water and slammed into the gallus frame; I jumped out of the wheelhouse to hook up the after door. When two-blocked (snugged up tight), each door was secured by a chain and hook that hung down from each gallus. The leader and then the legs were slowly wound onto the drums, bringing the mouth of the net ever closer to the side of the boat. At the helm, I put her into a slow circle to get to the downwind side of the trawl, allowing the *Dorothy & Betty II* to keep clear of all the twine that was billowing beneath the surface.

When she was laying just right, I threw the engine out of gear and stepped up to the rail just as the quarter ropes came up to the blocks on the gallus frames. I looked down into the water, brightened by our deck lights, and I could see a great number of big butterfish scattering in every direction as they worked their way through the top square of twine. There couldn't have been a better sight, as it was certain evidence of a good haul. We began whipping (tying off) the body of twine as quickly as possible to keep the fish within the net before sharks began circling the catch.

As the cod end came into sight, we could see that many of the butterfish were well above the splitting strap, so we took one split aboard, tripped it, set the cod end over again, and hardened the fish down into it. When we tripped the second bag, those big beautiful fish flooded across the deck. We had about seven thousand pounds of lovely silver and white large-count butters. Just as quickly as the men could move, we set the net over the side a second time, and I put the boat in gear and began streaming the net, legs, and leaders astern until they fetched up to the doors. These were quickly raised in the gallus frames, hooked up to the main towing cables, and released to run at the proper depth. Then we hooked the cables to the stern bitt and began the second tow.

This procedure was repeated twice more, but with each tow we took fewer and fewer fish. When the third one was on board, we kept the net on deck, brought the drag doors aboard, then pushed all the drag gear as close to the rail as we could get it. The butterfish were washed, run belowdecks into the pens, and iced down. The decks were washed clean, and the baskets and shovels and other loose equipment were all put down into the fish hold. The dory was brought down off the roof of the wheelhouse and put overboard to be towed astern. The kegs were put into place, as were the baskets of swordfish warp. The harpoon was secured across the striker's stand, the first dart was fitted onto the soft-iron shank (rod) at the end of the harpoon, and the warp was run aft to the first basket and keg. Now we would do what we all had been hoping to do when we left port.

First, we went below for a quick mug-up. Then Paul and Moe went aloft, and Norm climbed on top of the doghouse. I wanted him to be available on deck in case some problem arose. I gave Paul the course I wanted to hold. I planned to work outward from the channel edge where we had caught the butterfish into each point of an imaginary star.

Swordfish are almost always found in the same areas that attract butterfish because the big predators feed on them. As soon as the sun warmed the surface water, we were hoping to see a magnificent dorsal fin and tail riding high. And as sure as the tide ebbs and flows, about a half-hour after getting under way, we spotted the first fish. True to form, the men aloft screamed, "There he is!"

Paul steered us onto the swordfish, and without much ado I struck it. The keg went overboard and Moe went off in the dory to tend it. I just had time to retrieve the harpoon, place another dart on the shank, and tuck the warp into the leather toggles when Norman called out, "There he is!" Paul immediately turned the boat to put the sun behind us, thereby keeping the reflection off the water and allowing me to see underwater as we approached the fish. I drove the iron solidly through the widest part, and the fish dove, taking the warp and keg overboard in no time. I called to Norm to take a loran bearing on this one.

Seconds later, Paul roared, "There he is!"

"Where-away?" I asked.

"Astern of us, near the dory," he replied.

"Come about quickly," I called out as I again began to rig up the harpoon. If we got the fish, that would be three kegs out, and I yelled to Norm, "Get out another three kegs and get ready, just in case."

"Done, Cap. I have two ready now. Moe has his oar up. He must have that first fish alongside."

I shouted aloft, "Paul, if we iron this one, then once we're clear, let's quickly pick up Moe. Come on down to help Norm haul the fish aboard, then we'll tow Moe to the second fish. Norm, you go into the wheelhouse and I'll steer you with my pole." I instructed.

Because we were shorthanded and no longer had someone aloft to steer, the man in the wheelhouse had to take directions from the striker. The harpooner would point to the fish with his hand until the boat got close, then he would simply aim the pole in the right direction. As the moment of truth approached, the man in the wheelhouse would watch the striker's body language, and when he saw the pole thrown, he would quickly turn the boat to clear the warp.

We swung slowly over the swordfish as we were turning to starboard, putting me right over its back. I threw the pole, and the iron went in by the backbone. I could see the lily as it toggled beneath its belly.

"That fish is hit hard," I called over my shoulder to Paul as I again searched for a new lily and warp. "Tell Moe when we pick him up. He can haul as hard as he wants. He will never pull the lily out."

Coming alongside the dory, we picked up the strap that Moe had put around the tail of his fish. In one easy motion, Paul handed Moe a new strap for the next one. "Stay where you are, Moe. We have another fish ready for you and a third one that's alongside now. Sit down while we tow you over to the fish. Sam said to tell you that the third one is hit well and you should haul as hard as you can."

While Paul was talking, Norm came out of the wheelhouse and hoisted the fish aboard. It was a nice one that would likely dress out at well over two hundred pounds.

"Cap, what do you think about picking up the third one first?" Moe called out. "I can pull it quickly and then take my time on the second one."

"Good idea," I answered. "Norm, when everything is clear, tow Moe over to the fish we just struck."

This was a day that none of us would ever forget. I ironed fish after fish. We were running out of kegs, so I had Norm tie three life jackets together and use them as a keg. I then told him to set out a flag to use as a marker so we wouldn't drift off these productive grounds. We used a large bamboo pole surrounded with flotation material and topped with a cloth flag, and a large aluminum plate to serve as a radar reflector in case of fog. Norm weighted the bottom of the flag with a twenty-five-pound anchor.

Here's an example of how crazy things got: Moe had just picked up the third fish struck and we were steaming away from him. I had the harpoon in my hand and was about to rig it up. I had just put the lily on the shank and was running the warp up the pole prior to tucking it into the leather becket. Paul was at the wheel watching these preparations when he saw me quickly lift the pole and throw it. He thought I was kidding, but his experience took over and he turned the boat to starboard. As she swung, he saw the warp running overboard in quick order and watched as the life jackets went flying overboard.

The fish, a big one, had come up right under the stand as we were steaming away from the dory. I remember thinking that I wouldn't be able to keep enough strain on the warp to prevent the lily from dropping off the end of the shank. Fortunately, I managed to hold it tightly enough to drive the iron into the broad back of the fish. As I turned to pick up another rig, I saw that Moe had his oar up. He must have put his back into hauling that fish to the surface, as it took less than half an hour. If a doryman could retrieve a swordfish in under an hour, he was doing very well. I have hauled a well-hit broadbill in twenty minutes, but I wouldn't have tried that if it had been struck lightly. I once pulled a fish where the lily had penetrated just a little ways under the skin, missing all the vital organs. I was nearly two hours on that one. The only thing I could do was keep a light but constant strain on the warp, not giving the fish a minute of respite. Of course I was as weary as my quarry, but if I had let up, we might well have lost that broadbill.

By day's end we had harpooned and recovered fifteen swordfish. They were stacked up all over the deck by late afternoon. When it got to be 3:00 P.M., I walked aft to the wheelhouse and threw the engine out of gear. I told the men to haul the dory up on deck and clear the swordfishing gear off the working area. They looked at me in disbelief. "You're setting the net?" they asked in unison.

"Yes, I am. Now let's get moving," I answered, not allowing any room for argument.

Grudgingly, the crew hauled the doors up in their frames and swung them over the rail. In a moment the net was set, and I started around in a slow circle until I was heading in the right direction for the tow that I wanted to make. In forty-five minutes we hauled back and had about three thousand pounds of big-count butterfish, the same size as the ones we'd taken that morning, which seemed to have been days ago. Again we set out, and this tow yielded seven thousand pounds of the silver-sided darlings that I yearned to catch.

We then made our third and last tow, and I hauled back on the west end of it, bringing us that much closer to home. This haul needed to be split, and again the size of the fish meant that they would yield a good price on the New York market the morning after we took out at the Fishermen's Co-op in Galilee.

Following that final tow of the day, I threw the engine out of gear and we drifted for a while, allowing the crew time to clear the deck of butterfish and ice them down in the fish hold. When that was done, we put all of the nonessential gear belowdecks, where it would be out of the way. We then brought the drag doors aboard, chocking them tightly between the gallus frames and the bulwarks. Next we hauled the twine aloft, shaking out all the remaining fish and noting any tears that needed mending. As we lowered the net, we chocked it close in to the starboard rail, keeping the twine out of our way for the next project—dressing out the fifteen swordfish.

I sent Moe below to begin cooking supper. We headed west toward Point Judith and into the freshening sou'west breeze that usually made up on summer evenings. The *Dorothy & Betty II* slowly gained speed as I gradually increased the throttle to the rusty notch. Carrying the load that we had aboard on this day, she acted more like an island than a boat. She never lifted at all, and no little spray blew over the flare in the bow, landing back on the men as they dressed out those beautiful swordfish.

We lifted one fish at a time over the center checker, gutting it, then cutting lightly down both sides of the spine to remove the membrane that was full of congealed blood. This could be slowly pulled out in a single strip and disposed of. The initial bleeding had already been done while the broadbill was alive and lying alongside the dory, allowing the fish to pump much of its own blood into the water before being picked up by the mother ship.

By contrast, a swordfish caught on a longline, as are most of those on the market today, has retained all of the blood in its body because it has drowned while hooked on the line. When one of those fish is cut open, the flesh often looks grayish and almost muddy. But the flesh of a properly bled harpooned swordfish is bright and clean and pinkish when you slice into it. Most people think that the flavor of a harpooned fish is better, as well.

We had dressed out about five fish and iced them right on deck, covering them with a tarp to shed the spray coming over the rail, when Moe announced that supper was ready. I sent Norm and Paul below, then Moe came up to relieve me. When we had finished eating and Moe had cleaned up the dishes and secured everything below, he came up to help us finish the chore of dressing out the rest of the swordfish. All of them had been cleaned, iced down, and covered with tarps when the lights of Point Judith came into sight.

When we weighed out the next day, we had thirty thousand pounds of butterfish and over three thousand pounds of swordfish. Together, they made this one of the best stocks ever recorded by the *Dorothy & Betty II* in a single day. As we weighed out our trip, I thought back on my decision to stop swordfishing and set the net again. I knew that the crew had not been happy with my choice.

I had watched the three men very closely while we were towing. They didn't go aloft, but they could get a good view of the sea from the forward deck and the top of the doghouse that led to the fo'c'sle. Although swordfish had been horned out all day in nearly every direction, not one had finned after we set the net out. If we had chased them for the balance of the daylight, we would have missed out on upwards of twenty thousand pounds of butterfish.

It was this kind of fishing that I counted on to cover the hefty monthly boat payments and the wages for all of us on deck. I was as reluctant as the crew when it came time to stop swordfishing, but I knew better than to continue, and fortunately I have always been able to make the sound but boring decision when it was necessary.

That said, swordfishing is the most exciting thing a man can do and survive to do again.

It really is.

Captain's Log

Section IV
Epilogue

EPILOGUE

Barratry: A fraudulent breach of duty on the part of a master of a ship or of the mariners to the injury of the owner of the ship or cargo.

Webster's Ninth New Collegiate Dictionary

As stated in the opening chapter of this book, commercial fishing vessels are registered with the United States government. Once registered, usually at a U.S. Customs House, they technically became the property of the government. In the event of a marine casualty, (i.e., the sinking or destruction of a registered vessel), the captain/owner is responsible to report in detail how the loss occurred and to provide, in particular, information on the welfare and location of the officers and crew at the time of the casualty.

Early in the summer of 1962 an unauthorized person boarded the *Dorothy & Betty II* as she lay at our assigned berth in Galilee Harbor. We had just returned from a trip offshore and had unloaded our catch at the Point Judith Fishermen's Cooperative. After taking out the fish, we carefully checked the net, mending any tears; shook out all fish parts and trash that were hung up in the twine; then hoisted the net up off the deck so that it could dry.

The cook had returned from Stubby Caswell's market with all the grub needed for our next trip, which we had planned for early the next morning. The deck crew had finished their work, and the engineer had changed the oil and filters, and put new belts on the pumps and generators where needed. All the men left the boat at about the same time—just before noon. As he departed, the engineer made a final check of the deck hoses to be certain they were all coiled up, then he jumped up on the dock and ran to catch up with the rest of the crew. They were heading up to the Bon View for lunch and cold beer. The last to leave the *Dorothy & Betty II* was the cook, who had stowed all the food, iced down the perishables, checked the stove, shut off all the lights belowdecks, and then locked the doghouse companionway behind him. He, too, took a quick look overboard and on deck to see that everything was shipshape. Then he walked briskly up the dock toward his pickup, as he planned to meet the guys at the Bon View.

I had been at an unscheduled directors' meeting of the Fishermen's Co-op and was just getting into my truck as the crew started by. I waved them down to check on what they had done before leaving the boat. They assured me that everything needing repair had been cared for and that she was ready to sail on schedule. I told them I was headed home to have lunch with my family and would call them after the 11:20 P.M. weather report that night. If the forecast was good for the next few days, we would be leaving for a trip down east at midnight.

We did not sail as scheduled. After dark that night, someone climbed aboard the boat, jimmied the lock, and went down into the fo'c'sle. With grim determination, the intruder set the controls on the stove so that within a few minutes it would be flooded with oil and a fire would start.

That's exactly what would have happened if the ship's cook had not come down with some extra food supplies that he had forgotten earlier in the day. As he opened the fo'c'sle door and air rushed down below, it caused an explosion that belched smoke and fire up the ladderway and into the cook's face. The pressure knocked him over and left him disoriented. Once he sorted things out in his mind, he got up, ran to the nearest land phone along the bulkhead, and called both the fire and police departments.

Once the desk sergeant contacted me with the news, I phoned the rest of the crew to advise them that the trip had been canceled. I also asked them to come down to the boat to answer some questions for investigators. In the meantime, the police had called the Providence office of the FBI, which said that an agent should be dockside within the hour.

After a very thorough effort, investigators could not find enough evidence to determine just who had caused the damage. It was at this time that I was advised about the charge of "barratry" that was being applied as the basis for further investigation. The police, as well as the FBI agent, said that I could have the damage repaired and could go out fishing as soon as I wanted to.

I drew up a repair schedule assigning the crew to specific items that needed to be cleaned or replaced. I also called the owner of the stove shop and requested his involvement at the earliest time possible. Shortly after daylight the next day, the crew assembled and began their work.

By the end of the first day, all of the heavy damage had been repaired. Since space in the fo'c'sle was limited, only two men could reasonably do needed painting and varnishing, so further assignments were scheduled for the next day. The remaining man would work on deck or relieve the men below, as needed. Final repairs were quickly made, and by the third day all the interior painting was dry enough to allow us to schedule our next trip.

We quickly fell into our fishing routine, and in no time, two months flashed by and the crew had pretty well forgotten the explosion and fire. I hadn't heard anything from the police or FBI, so the whole event began to fade into the back of my mind, too.

July was only a week away when we decided that after the next trip, we would put on the swordfish stand and all the associated gear. Frequently swordfish would show up south of Block Island at this time of year, and the first one offloaded at the Co-op would bring one dollar a pound. A price like that, along with the fun of being the first boat to land a swordfish, made it worth the effort required to rerig the boat.

After one long day of work, the phone was ringing as my wife opened the back door and I struggled up the stairs with two sleepy kids. We had been visiting the in-laws and had stayed longer than usual, as we hadn't seen them in a couple of months. Gloria called to me, "Pick up the phone; it's the harbormaster in Galilee." That was not a good sign.

"Hey, Cap," he said. "We have a problem. Your boat has sunk at the dock. I've called the Coast Guard and the police. I don't know how you want to raise her."

Although stunned, I managed to say, "Thanks. I'll call the crew and be right down. Just don't let anyone go aboard until I get there." I turned to face my wife and told her about the sinking; Gloria immediately came to the same conclusion that I had: Somebody was deliberately creating problems for us.

"Why?" she asked. "Why is someone doing these things? Who could it be?" she asked.

"I don't know yet, but we're going to find out," I replied. "I'm going down to the boat. Don't wait up for me; I'll probably be all night." Gloria had already begun to put the children to bed before I got down the stairs.

I called the crew and asked them to come down to the boat. "It's likely to be an all-nighter, so bring your gear." Like most fishermen, we always kept a sea bag filled with clothing, shaving gear, and personal items—enough to sustain us for a few days—packed and within easy reach for any emergency.

When I arrived at the berth assigned to the *Dorothy & Betty II*, the docks were lit up like New York City. Two Coast Guard picket boats were alongside the flooded hull, in the same slip. They had already moved the other draggers that had been tied there. In addition to providing the picket boats with their spotlights and deck lights, the Coast Guard had backed some trucks up to the bulkhead, and sticking out the back end of these vehicles were banks of lights, all trained on the *Dorothy & Betty II* as she lay on her side. All the police cruisers that had formed a semicircle around the dock had their red, blue, and white lights flashing. It all looked like a scene from a Hollywood action movie, except my poor vessel lay on her starboard side like a struck and dying whale.

My brother John had moved his vessel, the *Four J's*, parallel to the *Dorothy & Betty II*, but across the slip at the next dock. This would give him a perfect angle to use his winch and cable should we need him.

After a cursory inspection of the scene, the Coast Guard officer-in-charge and I decided that it would be wise to wait until daylight to proceed with our salvage operation, primarily for safety's sake. It certainly was not worth risking the death or serious injury of my crew or that of the rescue team. Moreover, any damage to the electronics and the main engine had already been done, so a few hours of delay would not make things appreciably worse.

Daylight the next morning bode well for good weather and, hence, for a greater chance of success in our salvage project. The parking lot near the dock where the *Dorothy & Betty II* lay on her side was packed with cars, trucks, and curious onlookers.

A couple of navy divers put on their gear, first for a venture into the engine room and then into other watertight compartments, to ascertain

When someone sabotaged the Dorothy & Betty II *at her dock, the dragger ended up half-submerged and listing badly to starboard.*

Efforts to pull the vessel upright by attaching a cable to her mast were a disaster, as the mast and its supporting halyards snapped and destroyed virtually everything on deck.

any structural problems and/or the site of the ingress of water. Within the hour they reported no evidence of damage to the hull. The only difficulty we now faced was getting the dragger to sit upright in the water, and several plans were proposed. I suggested that we run a cable from my brother's deck winch through some fairlead blocks and across to the top of the *Dorothy & Betty II*'s mast. I reasoned that if a gradual strain could be applied as pumps expelled the water in the hull, then the boat should slowly right herself and rise in the water. Some engineer types suggested that the best location for securing the cable from the *Four J's* was just below our crosstrees, where the mast began to thicken, allowing more pressure to be applied.

In short order the cable was secured, whereupon I gave instructions that a very light strain be taken, allowing the hull to shed water and slowly right herself. The warning "very light" was sounded several times so there would be no misunderstanding. On my signal, John began engaging the winch.

What happened then, no one is sure of to this day. It appears that the winch jammed and continued to pull, putting a severe strain on the *Dorothy & Betty II*'s mast. Before the pressure could be released, the stays (cables supporting the mast) began to snap, and in an instant, that mighty spar from the forests of Maine cracked at the deck line. Down it came, destroying the companionway that led to the fo'c'sle, tearing the forward gallus frame from its heavily timbered frame, pulling the after mast forward and downward, and yanking the dory from its chocks atop the wheelhouse.

In a matter of seconds, more than fifty thousand dollars' worth of damage was done on the deck alone. In addition, all the antennae for electronics such as the radar, radios (three), and radio direction finder were ripped off the wheelhouse roof. One good thing did happen when our mast was torn from the deck: the hull turned level and upright. This allowed the Coast Guard to place three five-inch pumps inside the hull and begin removing water in a steady torrent, which allowed the boat to rise rapidly.

As soon as we could determine for certain that no new seawater was entering the hull, we knew that the dragger was in no further danger of sinking. As the vessel emerged from the water, my crew—along with

Coast Guard personnel—began a careful search of the *Dorothy & Betty II*'s deck, looking from stem to stern for the source of the seawater that had sunk her. In just a few minutes, they called me over to a spot on the starboard side, just aft of the winch. Our deck hose was lying over the caprail of the bulwark, and a layer of net twine had been flaked back and forth on top of it in an effort to hide it. We always coiled that hose on deck at the end of a trip, because leaving it overboard risked its siphoning water from the sea and back-flooding the engine compartment. I told the men to leave the hose as it was for a few minutes. I walked forward to the Coast Guard officer and asked him to come with me and then called out to my engineer to join us. The three of us walked aft to where the hose still lay partially covered as it went over the rail.

"Paul, as the engineer, what is your procedure when we shut everything down before leaving the boat?" I asked.

"The very last thing that I do is to haul the hose inboard," he responded. "Then I coil it up right in front of the winch housing, where the hose comes out from the engine room."

"Did you do that last night when we left the boat?" I asked.

"Yes, Cap, I sure did," he answered.

Norman, my first mate spoke up: "I was working right alongside Paul, and I watched him coil that hose, just as he always does."

"Thanks, Norm." I said, as I turned toward the officer. "This was not an act of forgetfulness. This was a deliberate act of sabotage, and I think you should contact the FBI."

"I agree," said the officer. "I'll phone the agent right now so he can get down here to see the damage before you begin to clean up."

My wife called to me from the dock, and when I went over to her, she said, "I think you ought to talk to your brother. He's really shook up about the mast." I went across the dock to John's boat, where he was leaning back against the front of the wheelhouse. His face was ashen, and I knew that he felt some responsibility for all the damage.

"Hey, Slag, ease up on it. This was an accident and no one's faul," I said. "You did everything by the book. The winch stuck in gear. We've both had it happen before, and we'll probably see it again. You made a good effort, and I thank you for trying. I know you're anxious to get to sea again, but could you tow us over to Jamestown? I want to haul out at the Round House Shipyard."

"Sure!" said John. "When do you want to go?"

"As soon as the Feds release us, I guess. They should be down within the hour to have a look. There's no reason to hold us up as far as I can see."

"Give me a call as soon as you hear," said my brother. "We can swing off to the east'ard from Jamestown just as well as we can from here." John spoke quietly, which wasn't like him at all.

"Okay," I responded. "I'll call you as soon as we get clearance. Right now I need to take a break; I'm wound right up."

It was almost 4:00 P.M. before the Coast Guard and the FBI finished their inspection and released my boat to be moved to the shipyard in Jamestown. I phoned my brother to ask when he wanted me and my crew to be ready for the tow.

"We're all set to head out, so it will just take about an hour to round up the crew," said John. "Let's say 5:30. Since we'll have fair wind and a fair tide, it'll only take us two hours to get to the shipyard. Does Earl Clark know you'll be coming over?"

"Yes," I answered. "He said that the cradle is already down and ready for us, and he is going to keep the yard crew available until we get there."

"Okay," he grunted. "See you in a little bit," and he hung up.

I called my crew and told them to head down to the dock. I asked Gloria to arrange for our transportation home from Jamestown. Some of the other wives had vans that could hold most, if not all of us. Soon we were heading out of the breachway on our way to the Round House Shipyard. It would be nearly a year before we would be in-bound through that self-same breachway with a new mainmast, after mast, rigging, doghouse, turtleback, and dory rack. Thousands of dollars' worth of new electronics had been installed, the hull was freshly painted (inside and out), and there was a glistening paint-and-varnish job in the wheelhouse and fo'c'sle.

On the day of our return, the *Dorothy & Betty II* looked as pretty as the day I'd bought her up in Maine. The only things lacking were a new generator and an overhaul of the main engine. That's why we headed straight for Ken Gallup's repair shop. We were in for a long haul of dismantling, cleaning, and rebuilding our main source of power, that screaming hunk of cast iron and other sundry parts that we affectionately called our "Jimmy"—or, more correctly, our General Motors diesel. Many days of summer heat, cussing, and bruised body parts concluded in an engine that roared to life with the trademark GM scream.

We were nearing our goal, which was casting our nets once again into the stream of life, the ocean blue. What a long, difficult wait it had been, and we were no nearer to identifying the culprit responsible for all of the pain, the aggregation, the tens of thousands of dollars in repair costs, and the even greater amount of money we had lost by not being able to fish. And the injury was not to me alone, although I was clearly the prime target of the "vandalism." My crew and their families paid dearly for the warped mind of some individual who was out to destroy me financially. He had come very close to being successful.

At last came the day for sea trials, and what a happy day it was. The main engine was running like a watch—a big watch, but a watch just the same. All of the ancillary power plants, pumps, and machinery had been fine-tuned and were humming in harmony. Ken Gallup, owner of the repair shop and its chief mechanic, had just advised me that everything was ready for a trial run. He and another mechanic were to go to sea with us to be certain that they had done their jobs properly.

All the dock lines were tossed free of the pilings and coiled on deck. As we backed out of our berth, I sounded the air horn in celebration of this momentous event. A few mechanics on the dock gave us a mock cheer, as though we were an ocean liner leaving the harbor.

Free of land at last, I put her in forward gear and moved the throttle up just enough to hold our headway, as I was determined to break in the engine very gently. It was not until we were outside of the walls of the Harbor of Refuge that Ken instructed me to increase our speed gradually until he told me to hold the engine at that rpm. We steamed into the sou'west chop that was making up, and the old girl would raise gracefully and then settle gently into the next oncoming sea, just as she always had. What a delightful feeling. It had been nearly a year since the deck beneath my feet had moved so gracefully.

Suddenly the big diesel slowed down almost to a stall. I ran aft toward the engine-room companionway just as Ken stuck his head out and hollered at me to call the Coast Guard and to turn the boat around and head in. He also told me to send down a couple of crew members—quickly. I did as Ken requested without taking the time to ask him why such actions were necessary.

In a few minutes, he climbed out of the engine room and into the wheelhouse to tell me what had just happened. Ken said that the engine had been running just fine and that he had been about to request full speed when he noticed water beginning to seep in around the seacock. He and his mechanic immediately climbed over the engine and sundry other mechanical devices and got alongside the through-hull just as it began to blow inboard, along with a great volume of water. Ken was able to throw his body against the seacock as his assistant re-threaded one of the bolts holding the whole piece of hardware together. With that done, they were able to fasten the other bolts just as they, too, were about to give way from water pressure.

With a slight quiver in his voice Ken said, "I secured that seacock myself, and George, the feller with me now, worked alongside me as I did it. This was no accident, Sam. This is out-and-out sabotage. When you radio the Coast Guard, you ought to ask that the FBI meet us, as well. And I want to talk to them, too."

I called the Coast Guard on its regular channel and asked to speak to the officer on duty. I told him briefly about our incident and asked that they again contact the FBI in Providence. This done, I concentrated on bringing the *Dorothy and Betty II* into the harbor and then to the dock. Ken and his assistant headed back down into the engine room, and I told two of my men to join them.

As soon as the FBI agent and the Coast Guard watch commander arrived at Ken's place, we secured the boat and cleared the deck of any fishing gear that might be in the way of those carrying out the investigation. I had the cook spruce up the galley, make a couple of pots of coffee, and break out some pastry. We would be conducting interviews for quite a while, and I decided that we might as well be comfortable while doing so.

When the questioning began, the inquiries were directed to everyone in general: Where were you and when? Instead of asking members of the crew specific questions, they were invited to offer their opinions. I thought that was a strange way of conducting the investigation, but I wasn't the expert in charge. When the first session was over and everyone had downed some coffee and goodies, the crew was invited to go topside but was told to stay aboard the boat. Each man was going to be called in individually when needed. I didn't notice it at first, but a couple of the Coast Guardsmen had sidearms, and one stood leisurely by the companionway to the fo'c'sle, while the other one positioned himself by the engine-room hatchway.

After the inquiry went on for a couple of hours, we were told that we could go home if we wanted to. The only conclusion seemed to be that all three incidents—the fire in the fo'c'sle stove, the sinking at the dock, and now the tampering with the seacock—were part of a pattern that meant something to the investigators. It was at this time that the charge of "barratry" was again applied to the case.

My crew and I took a day off to catch our breaths and then began our regular schedule of fishing. We did allot some time to restore the mind and the soul—we went after swordfish. It was July, so we knew they would be showing up south of Block Island, and there is nothing more refreshing than chasing those old broadbills. Although I enjoyed a summer doing the things that I loved, some ominous pressure was building inside me. As much as I wanted to cast aside recent events—and especially to forget that someone had such hostile feelings toward me—my body was reacting to the stress.

I was at a business meeting ashore when I began to feel as if I was coming down with the flu, and as the physical distress increased, I had to excuse myself from the proceedings. As I was driving home, I realized that I was suffering not from the flu but from a heart attack, a conclusion I reached when my left arm went numb. I immediately faced a critical decision: Do I continue to drive home and possibly kill someone else when I pass out, or do I pull the car to the side of the road and die alone, out in the wilderness? I chose to continue driving but with great trepidation. Fortunately, I made it safely to my doctor's office. After a quick exam, I was rushed to the local hospital, where I was confined for six weeks.

I guess you know what happened next. My doctor, a cardiologist, was aware of the ongoing attempts to destroy the *Dorothy & Betty II,* and very likely, me, as well. I believed that I could cope with that problem; what I could not handle was the danger to my crew that resulted from the efforts of this saboteur.

The third attempt—the loosening of the bolts on the seacock—was no mere nuisance. Had it happened a half-hour farther offshore, or if it had gone undetected, the boat could have sunk with all hands before we received assistance. The flood of water would have caused the engine to stop, shorting out the batteries powering the radio needed to call for assistance.

I knew that I could not be responsible for the potential loss of my crew because some psycho had an illusion of some kind about me. If I had to give up my lifelong dream of owning and fishing my own dragger, I would do that to protect my crew. These were not just men who worked *for* me; they worked *with* me. They were my friends, and some were my spiritual brothers. I could not—I would not—endanger them just to satisfy my own wishes and dreams. The decision was mine alone to make. I agonized over it for weeks.

I finally made my choice: I came ashore. Permanently. My son was shocked, as he had counted on one day running the *Dorothy & Betty II,* in the tradition of all fishing families, where the eldest son inherits his father's boat, and, in time, passes that gift on to his own son. In my family, this tradition was broken.

To my knowledge, my beloved *Dorothy & Betty II* is still fishing out of Portland, Maine. At least that was where I saw her last.